The Evolution of
VERTEBRATE
DESIGN

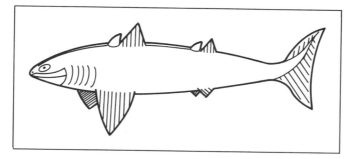

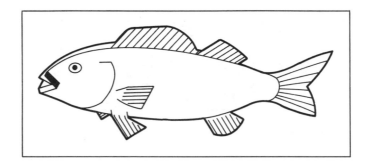

The Evolution of
VERTEBRATE
DESIGN

Leonard B. Radinsky

The University of Chicago Press
Chicago and London

The University of Chicago Press, Chicago 60637
The University of Chicago Press, Ltd., London

02 01 00 99 98 97 96 95 94 93   5 6 7 8 9 10

Library of Congress Cataloging-in-Publication Data

Radinsky, Leonard B.
    The evolution of vertebrate design.

    Bibliography: p.
    Includes index.
    1. Vertebrates—Evolution. 2. Vertebrates,
Fossil. I. Title.
QL607.5.R33 1987   596'.038   87-5959
ISBN 0-226-70236-7 (paper)

⊗ The paper used in this publication meets the minimum requirements of the American
National Standard for Information Sciences—Permanence of Paper for Printed Library
Materials, ANSI Z39.48-1984.

To Adam, Josh, and Leah

Contents

Publisher's Note

Leonard B. Radinsky was born in 1937, the son of Russian-Polish immigrants, in Staten Island, New York. He developed a keen love of fossils as a young boy and at the age of thirteen produced a short "book" illustrated by himself, entitled "Prehistoric Animals." He did undergraduate work at Cornell University and received the masters and doctorate degrees from Yale University. After a postdoctoral fellowship at the American Museum of Natural History, he taught at Boston University and Brooklyn College and then in 1967 joined the University of Chicago faculty.

Leonard Radinsky died of cancer in August 1985. At the time of his death, he was professor and former chairperson in the Department of Anatomy at the University of Chicago. During his distinguished career in biology, he published numerous scientific and popular articles, the best known of which are those on the neuroanatomy of extinct mammals and the functional morphology of carnivores.

Among friends, Len's love of teaching is especially well remembered. *The Evolution of Vertebrate Design* was written as an outgrowth of his course on vertebrate morphology, paleontology, and evolution for undergraduates not majoring in biology in the College of the University of Chicago. He wanted his students to appreciate the evolution of vertebrates through an understanding of how their body form, or design, changed through evolutionary time. Finding no text that combined information on functional morphology and paleontology, he decided to write this book himself. He was in the process of revising the manuscript when he died. The work of completing the book was done by Sharon B. Emerson. The Press is very pleased to publish *The Evolution of Vertebrate Design,* and we share its author's hope that it will bring some of the excitement of the evolution of vertebrate life on earth to the students who read it.

Acknowledgements ───────

There are many people who helped Len with his book. Harry Greene, Lynne Houck, Jim Hopson, Keith Thomson, John Bolt, George Lauder, and Mike LaBarbera all read early versions of various chapters. Their comments and criticisms were greatly appreciated. Jack Sepkoski, Philip Gingerich, Jacques Gauthier, Jim Hopson, Eric Lombard, and George Lauder contributed comments and data for figures. Discussions with Martin Feder, George Lauder, Jim Hopson, and Eric Lombard were particularly helpful to me. Dennis Green did a wonderful job of translating Len's ideas and sketches into finished figures. Ruth Bachman compiled the glossary and helped read proof.

Susan Abrams took a special interest in Len's book from the beginning. Her strong emotional and professional support are largely responsible for the book's completion. Finally, Len would certainly have wanted to acknowledge the love and encouragement of his friends. And so, a special thanks to Lynne Houck, Steve Arnold, Harry Greene, Kristine Tollestrup, Jeanne and Stuart Altmann, Susan and Brock Cole, Steve Gould, Ann Sakai, Steve Weller, Brigid Hogan, Martha McClintock, Pam Parker, and Wim Weijs.

Sharon Emerson

The Evolution of
VERTEBRATE
DESIGN

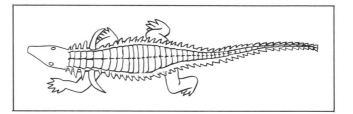

Introduction: Fossils and Phylogeny

CHAPTER 1

The history of vertebrate life is a fascinating story, one that chronicles the rise and fall of numerous groups of organisms, including strange and bizarre types like nothing alive in the world today. It is a story of major evolutionary transformations over millions of years, of long periods of little or no change, and of disastrous worldwide extinctions and great evolutionary radiations. And, of course, the story includes our own evolutionary history, which can be traced from our early ancestors among ancient jawless fish. This book will look at the important changes in basic body organization that have occurred over the past 500 million years and at the periodic evolutionary radiations that produced a stunning diversity of life forms at various points in time. Emphasis is placed on explaining the functional significance of evolutionary changes in anatomical structure. We will look at the evidence and methods of analysis used to obtain those explanations, so that you can see how we have come to our current understanding of ancient life forms.

Sources of Information

We have two main sources for the evolutionary history of the vertebrates: the fossil record and the diversity of living vertebrates. The fossil record is very incomplete, for it preserves evidence of only a very small percentage of all the species that ever existed, and it usually provides information only on their hard parts or skeletons. However, it is our only source of information about extinct species. The fossil record is set in the framework of geological time and therefore can provide direct evidence of actual evolutionary transformations at fairly specific periods through time. Furthermore, with careful analysis, skeletal remains can provide much insight into the ways of life of extinct animals.

Fossils are the traces or remains of extinct organisms, sometimes just imprints in rocks that were once soft sediments but often the actual mineralized remains of the hard parts of the body. Two questions commonly asked about fossils are, How they are found? and How they are dated? Most fossil localities are found more or less by accident—in the course of geological explorations or commercial excavations or stumbled upon by someone who notices an unusual looking bone lying on the ground. Once they are discovered, rich fossil localities are usually visited time and again by paleontologists. Some of the most productive localities today are places that were first discovered in the middle of the last century, and they have been worked on and off for over a hundred years. More rarely, fossil localities are found by paleontologists who deliberately explore regions where rocks of the desired age and of the right kind to preserve fossils are exposed at the surface, bare of vegetation and soil.

Fossils are dated by radiometric "clocks," that is, elements with unstable forms that change (decay) spontaneously and at a regular rate compared with other elements. The two most important clocks for fossil dating are uranium 238, which decays to lead 206, and potassium 40, which decays to argon 40, at known and very slow rates. When a rock solidifies from a liquid state (such as molten lava), its atoms are frozen into the lattices of mineral crystals and the decay products of unstable elements begin to accumulate next to the as yet unchanged atoms. Thus from the amount of decay product relative to unchanged parent element, and with a knowledge of the rate of decay, we can calculate the length of time that has passed since the rock solidified. Now, fossils are not preserved in rocks that solidified from a molten state (igneous rocks). Rather, they are found in sedimentary rocks that form from the compaction and cementation of layers of sediments, such as mud, silt, and sand. Such rocks can be dated only indirectly, by their position above or below igneous rocks that contain radiometric clocks. For example, the oldest known direct human ancestor, *Australopithecus afarensis,* is dated at between 3.8 and 3.6 million years old because it was found in sedimentary rocks sandwiched between lava flows of those dates.

Living species are our other source of information on the evolutionary history of vertebrates. They can provide a vast amount of data on anatomy, physiology, and behavior. These data can be used to infer the pathways of evolutionary transformation through geological time that resulted in the diversity of modern forms of life. A common approach to the study of evolutionary transformations begins with the selection of a group of living species assumed to represent an evolutionary se-

quence—for example, a fish, an amphibian, a reptile, and a mammal—
and to use that series to study the transformation of a given organ or
anatomical system of the body. We call such a series a *scala naturae,* or
"scale of nature," because it is assumed to represent an ascending sequence of organisms, from primitive to more advanced.

However, a potential problem with this approach is that all living
species, being end products of unique, individual evolutionary histories, are mosaics of primitive and advanced features. We cannot always
be sure that a particular feature in one living form represents an ancestral stage of that feature as seen in another living form. For example,
mammals evolved from reptiles, and modern reptiles are generally considered to be more primitive than modern mammals. Therefore it is
often assumed that modern reptiles represent an evolutionary stage that
the ancestral mammals passed through. However, although mammals
evolved from reptiles, they did not evolve from the sort of reptile that
exists today. Therefore, we would be misled if we considered the modern reptile brain, for example, to represent an early evolutionary stage
of the mammal brain. The forebrains of modern reptiles are dominated
by an advanced integrating center (called the dorsoventricular ridge)
that is not found in the brains of modern mammals. Instead, mammals
developed a different part of the forebrain (the cerebral cortex) as a
major integrating center. In dealing with presumed evolutionary sequences based on living species alone, we must always consider the
possibility of mosaic evolution and look for independent evidence that
the presumed primitive condition of the system under study is indeed
primitive.

A more productive approach to understanding historical transformations of morphology from living species is to compare changes between
forms that are most closely related. But first the relationships among the
organisms in question have to be established. Systematics is the special
discipline in biology that studies the evolutionary relationships among
organisms. Such genealogies or historical hypotheses of relationship are
called phylogenies. Phylogenies or evolutionary relationships of groups
that are presented in this book are constructed from a study of the distribution of primitive and derived (or advanced) character states.

For any character or trait of an animal, there are often several conditions or character states. For example, if the heart were a character
being studied in vertebrates, the different character states might be defined by how many chambers the heart has in different groups. All

character states have polarity; that is, one character state has evolved sequentially from another. It is the direction of this polarity that establishes which character state is primitive and which is derived. Using the heart example again, we are interested in establishing whether a four-chambered heart is primitive and a two-chambered heart is derived (or advanced), or vice versa. Sometimes a character state is considered primitive when it is the most widely distributed state among members of other groups of organisms than the ones being studied. This method of establishing polarity is called outgroup comparison. Another method for establishing polarity is to examine the ontogenetic development of the character. The adult character state that appears first during development is sometimes considered to represent a more primitive condition than a state appearing later. Using our heart example once again, if during development the heart starts out with two chambers and changes to four chambers later in ontogeny, the two-chambered heart might be considered primitive.

The polarities for a large number of characters are established for any phylogenetic analysis. Once the character states for each character have been determined to be either primitive or derived, the number of shared derived character states are used as the basis for determining degree of relationship among the organisms being studied. Those organisms with the greatest number of shared, derived character states are considered to be most closely related. They are placed adjacent to each other on a branching diagram.

As mentioned above, an important approach for reconstructing evolutionary transformations from living species is through the examination of ontogeny, or the course of an individual's embryonic development. The late nineteenth-century biologist Ernst Haeckel proposed as a "law" that ontogeny repeats or recapitulates phylogeny, meaning that the individual, in the course of its embryonic development (ontogeny), goes through the same stages that its ancestors did in their evolutionary history (phylogeny). Over the past century, this idea has been variously in and out of favor. Currently it is the subject of renewed interest, with the understanding that it is ancestral developmental stages rather than ancestral adult stages that can be seen in modern embryonic development. A classic example of the value of this approach is the demonstration from developmental series that two of the tiny bones in the middle ear of mammals (the malleus and the incus) were derived from the jaw bones (articular and quadrate) of our reptilian ancestors. That is, the early developmental stages of ear bones in mammals resemble early stages of jaw development in reptiles. This inferred evolutionary transformation of reptile jaw bones to mammal ear bones was later confirmed with the discovery of "missing link" fossils that display the intermediate condition in adult stages.

Of our two sources of evidence on the evolution of vertebrates, fossils preserve only a small amount of the total anatomy of the once-living animal but can be dated and placed in temporal sequences. It is important to remember, however, that the majority of extinct forms are likely to be a mosaic of primitive and derived features just like most living species. The geologically oldest representatives of a group do not necessarily have only primitive character states. For this reason, scientists also rely on embryonic data and outgroup comparison to establish which character states are primitive and which are advanced. Often the outgroups used for comparison include fossils, but living groups are part of the analysis as well.

Phylogenetic relationships inferred from outgroup comparison and embryonic data are independent of the time sequence known from collecting and dating fossils. The time-sequence relationships can then be used as an independent "check" or test of the proposed evolutionary sequence. Figure 1.1 shows a proposed phylogeny for mammals and mammal-like reptiles based on the analysis of 200 characters for which primitive and derived character states have been defined from outgroup

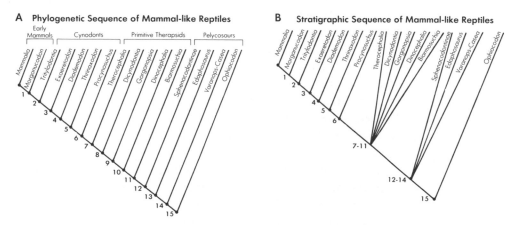

Fig. 1.1 *A*, phylogenetic relationships of mammal-like reptiles based on analysis of anatomical characters. *B*, phylogenetic relationships of mammal-like reptiles based on geological age of the specimens. (after J. Gauthier)

comparison and embryonic data. Figure 1.1 also shows a phylogenetic sequence of the same organisms based on the geological age of specimens. You will see that, in this case, the two diagrams are in good accord. The same evolutionary relationships have been obtained from the time sequence of fossils as from outgroup comparison and embryonic data on living and fossil species. Throughout this book, groups are referred to as primitive or advanced. These terms are applied on the basis of the relative number of primitive and derived or advanced character states found in the group and on the basis of examination of the time sequence of change from the fossil record.

Methods of Analysis

There are two main approaches to interpreting the functional significance of evolutionary changes and the ways of life of extinct species: form-function correlation and biomechanical design analysis. Form-function correlation involves looking for the behaviors or functions that are correlated with a particular anatomical form in living species and then extrapolating the correlation back to extinct species in order to infer function from their form. For example, fast-running animals today, such as horses, antelope, and deer, have relatively long legs and very long feet. From this association of long legs and speed in living animals, we can infer that extinct species with relatively long legs may well have been fast runners. The form-function correlation approach works well as long as we have living species with the anatomical features in question, but many fossil species display features unknown in the modern world, and in these cases we must use the second approach, biomechanical analysis.

Biomechanical design analysis involves looking at an anatomical structure from a biomechanical or engineering perspective and inferring from its shape and construction how well it would perform a given function. For example, one group of dinosaurs, the dome-headed dinosaurs, or pachycephalosaurs, had enormously thick and dense skull caps overlying their brains. There is nothing like that in living animals for comparison, but a biomechanical analysis of the shape of the skull shows that it meets design criteria for absorbing large forces or shocks from head-on blows. Given shock absorption as the probable function of the thickened skull caps, a likely interpretation of their significance with respect to biological role, or way of life, is that pachycephalosaurs lived in social groups and used head butting to establish dominance hierarchies, just as mountain goats and mountain sheep do today.

Sometimes we compare a particular morphology with a theoretical optimal design for a given function, or we compare different morphologies with each other and with an optimal design. Optimality is defined by the amount of energy it takes to perform a specific function. When these kinds of comparisons are made, we sometimes talk about the relative efficiency of a given design. Efficiency is a term that is used repeatedly in this book to discuss morphological transformations or changes. One structure is considered to be more efficient than another if it is closer to the biomechanically defined optimal design, that is, if it requires less energy to perform a given function. Energetic efficiency can be measured by comparing mechanical efficiency of a system (see chap. 7 for more detail) or by measuring the amount of oxygen it takes for two animals with different morphologies to perform a given function.

Ideally, form-function correlation and biomechanical design analysis approaches are used together to examine a given problem or question. Take, for example, the long legs and fast running correlation we have just considered. Biomechanically, one of the ways to increase running speed is to have a longer stride length. Larger steps mean that an animal will cover a greater distance for a given limb movement or expenditure of energy. The longer legs seen in fast runners are consistent with biomechanical predictions of how to increase speed.

Historically, these approaches to the functional analysis of form have not always been rigorously applied. As a result, early form-function studies often resulted in erroneous conclusions. For example, the earliest primates (the order of mammals including lemurs, monkeys, apes, and humans) had enlarged pointed incisor (front) teeth, superficially similar to those of modern rodents, such as squirrels and mice. The enlarged incisors of modern rodents seem superficially well suited for nibbling on seeds, nuts, and berries, and one can see squirrels and other rodents eating such food in parks and zoos. Therefore, for many years, biologists assumed that the enlarged incisor teeth of the earliest primates indicated that they fed on such plant food. The origin of the order Primates was therefore thought to be correlated with a shift in diet from the primitive insectivory to herbivory. However, a more careful, detailed analysis of the diets of many different rodent species revealed that many modern rodents are actually omnivores. They feed on insects and other animal material as well as on vegetation. Further, a recent form-function analysis of the correlation between diet and the size and shape of molar teeth in many living species showed that the type of molars seen in the earliest primates is correlated with a basically insectivorous diet. Thus the original, superficial analysis of form-function correlation that rodent incisor shape necessarily indicated an herbivorous diet proves incorrect. Our current understanding of the origin of the order Primates is that no shift in diet was involved and that, like their ancestors, the earliest primates were basically insectivorous. The moral of this story is that it is not enough to have superficial observations on possible form-function correlations or to make intuitively reasonable assumptions. We must carefully and rigorously gather data to establish form-function relationships.

Classification

The history of vertebrate life is a story of transformation and diversification, of origination and extinction, that produced an incredible variety of life forms. To help comprehend that diversity of organisms and their relationships to one another, we must use a system of classification. The system used almost universally today was invented about 200 years ago by a Swedish botanist named Carl Linneaus. The Linnean system uses a hierarchy of categories to reflect the degree of evolutionary, or phylogenetic, relationship: species, genus, family, order, class, phylum, and kingdom. Species are groups of organisms that in the wild can interbreed with each other and produce fertile offspring. Genera (the plural of the word *genus*) are groups of closely related species, families are groups of related genera, and so on. For example, dogs are classified as the species *familiaris,* in the genus *Canis* along with wolf, coyote, jackal, and other species. Species names are always written with their genus names, so the dog species is written *Canis familiaris,* sometimes abbreviated to *C. familiaris,* and genus and species names, because they are Latin or Latin based, are always written in italics or underlined. The genus *Canis* is a member of the family Canidae, along with red and gray foxes, African hunting dogs, and about a dozen other genera. The family Canidae is in the order Carnivora, along with about ten other families, including those of the cats, weasels, bears, and seals. The order Carnivora is one of about twenty orders of the class Mammalia, which includes the warm-blooded, suckling, furred vertebrates.

Intermediate categories are used in the Linnean system of classification, and vertebrates, the animals with backbones, are put in a category intermediate between class and phylum, the subphylum Vertebrata. Other classes in the Vertebrata include those of fishes, amphibians, reptiles, and birds. To finish the classification, vertebrates are included with a few other subphyla (phyla is the plural of *phylum*) in the phylum Chordata, the animals with a skeletal structure called the notochord, which we will learn more about in chapter 3.

We can look at humans as an example closer to home. We are the species *sapiens* of the genus *Homo*. *Homo sapiens* is the only living species in the genus, which includes also the extinct species *Homo erectus* of about .5 to 1.5 million years ago. The genus *Homo* is placed in the family Hominidae, along with the extinct genus *Australopithecus*, the earliest upright walkers, known from fossils dated to about 1 to 4 million years ago. The family Hominidae is included in the superfamily Hominoidea, along with the ape and gibbon families (Pongidae and Hylobatidae), and the superfamily Hominoidea is placed in the suborder Anthropoidea, along with the superfamilies of New World and Old World monkeys. Finally, the Anthropoidea is classified in the order Primates, along with the so-called lower primates, the lemurs, lorises, galagos, tarsiers, and some groups of early fossil primates. In descending order, then, humans would be classified as: phylum Chordata, subphylum Vertebrata, class Mammalia, order Primates, suborder Anthropoidea, superfamily Hominoidea, family Hominidae, genus *Homo*, species *Homo sapiens*.

Of all the categories in the Linnean system of classification, only the lowest, the species, has an objective definition and can be tested in a rigorous way: members of a given species must be capable of interbreeding and producing fertile offspring. In practice, although the species category is defined by this particular criterion, we recognize the members of a given species by the visible similarity among them. The members of a given species, despite some individual differences, tend generally to look like one another, and we do not often test species boundaries by the interbreeding criterion. Thus, in practice, both fossil and living species are recognized in similar ways. The higher groupings, genus and above, are based on the subjective opinions of the specialists who study particular groups of organisms, and they often have differences of opinion on how the higher categories should be arranged. In this book, phylogenies will be provided with each discussion of a major group of vertebrates. Because we are dealing with an overview of vertebrate evolution rather than a detailed description, our phylogenies will be primarily on the level of class and order.

Vertebrates in Geological Time

The history of vertebrate life is a history that covers about 500 million years, an immense time span, although only a short part of the age of the Earth, which is about 4.6 billion years old. Most accounts of vertebrate evolution are set in the context of the geological time scale (fig. 1.2), and events are dated in terms of millions of years. While most of us have a good feeling for what time means in terms of days, months, and years, it is hard to comprehend the passage of time of thousands or millions or years. Therefore, to provide a time context for vertebrate evolution that will be more easily comprehensible to us, we can translate the geological time scale of millions of years into a scale in which the age of the Earth is one year. We can then place the major events of vertebrate history in this relative time scale of a year. If we consider the Earth to have condensed as a planet on January 1 of a time-scale year, then fossil bacteria, the oldest evidence of life on Earth (about 3.8 bil-

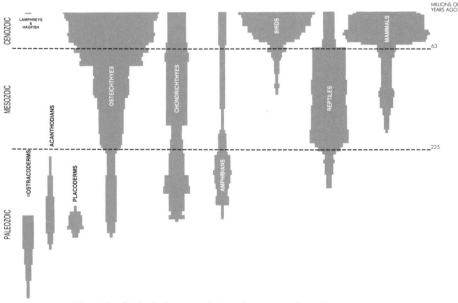

Fig. 1.2   Geological ranges of the major groups of vertebrates.

lion years ago), would have existed on April 4. The fossil record is very poor through the spring and summer of this year, and not until November 9 (about 650 million years ago) do several different groups of marine invertebrates evolve hard parts, and we begin to get a good record of many different phyla.

The oldest record of vertebrate life appears on November 20 (about 520 million years ago), and we have a good fossil record of many kinds of jawless fishes by November 29. Jawed fishes evolve on November 30 and undergo great evolutionary radiations during the month of December. The first land vertebrates (primitive amphibians) appear on December 3 (360 million years ago), and the earliest reptiles on December 7 (310 million years ago). Between December 9 and December 23, there are extensive evolutionary radiations of reptiles that produce the dinosaurs, flying and swimming reptiles, and the first birds and mammals (fig. 1.2).

Mammals first appear on December 16 but are small, insignificant animals until late on December 26. Then, after the extinction of most of the reptiles, they undergo an explosive evolutionary radiation that in the last four days of the year results in most of the different kinds of mammals we see on Earth today, as well as several extinct groups. Our own order, the Primates, appears in that radiation on December 26 (65 million years ago), but it is not until early on December 29 (30 million years ago) that higher primates (the first monkeys and apes) evolve. The oldest hominids (*Australopithecus* species) appear around 4 P.M. on December 31, and the oldest *Homo sapiens* sometime between 11:30 and 11:45 P.M. on the last day of the year.

Now that we have a comprehensible time framework, we can take a closer look at what happened during the last 41 days of our time-scale year, during the 500-million-year evolutionary history of the vertebrates. We can begin with the basic evolutionary processes that have produced the great array of vertebrate life on Earth, and this is the subject of chapter 2.

Evolution: Patterns and Process

As we have seen, the fossil record of vertebrate evolution, with its intermediate forms and transitional series set in the framework of geological time, provides factual evidence of evolution. But this record does not tell us how, by what process, evolution occurs. Our understanding of how evolution works is based on studies of living species. The main outlines of the current theory or model of the process of evolution were worked out in the mid-nineteenth century independently by Charles Darwin and Alfred Russel Wallace.

The Process of Evolution

A major question for nineteenth-century biologists was how the enormous diversity of living species had come about. After he had spent twenty-five years studying many different kinds of living organisms, Darwin came up with three facts from which he developed a theory for how evolutionary transformations (descent with modification) might come about. What were Darwin's facts? First, there are differences between individuals of the same species. Second, offspring tend to resemble their parents. Third, more individuals are born than can possibly survive. The last observation means that there will be differences in reproductive success, or fitness, among individuals of a species. Some individuals will not survive to sexual maturity. Among mature animals, some individuals will produce more offspring than others. The factors that result in some individuals dying before reproducing (or reproducing less than other individuals) Darwin called natural selection. The great insight of Darwin and Wallace was that forces of natural selection, operating on genetically controlled variation among individuals, could result in changes within lineages over time, or evolution.

Darwin recognized that evolutionary change through natural selection would take a very long time and that therefore one could not expect

15

to observe evolution directly. Consideration of the diversity seen among breeds of domestic plants and animals, however, indicated that strong artificial selection could produce within relatively few generations differences similar to those seen among wild species and genera. Thus evolutionary change by natural selection seemed a plausible theory.

What Darwin and Wallace did not know was how characteristics or traits of organisms were transmitted from parents to offspring. Our modern understanding of how genes and chromosomes work came first from plant and animal breeding experiments in the early part of this century. This early work resulted in theoretical models of genetic transmission of characteristics. Breakthroughs in molecular biology in the past few decades have provided insights into how genes actually work at the molecular level. The observable features of an organism—behavior, physiology, and morphology—comprise its phenotype. The particular combination of an organism's genes make up its genotype. Genes are complex molecules composed of long sequences of building blocks called nucleic acids. The nucleic acids are arranged in very long, spirally twisted double chains (the famous "double helix"). The specific sequence of nucleic acids in a gene provides a code for the manufacture of specific proteins and enzymes needed for the construction and maintenance of an organism's phenotype. Large numbers of genes are linked together in structures called chromosomes. In sexually reproducing organisms, half of an individual's genes come from one parent and the other half from the other. Thus each fertilized egg contains a new combination of genes, different from those of each parent, and this recombination of genes in sexual reproduction is one source of genetic variation among individuals of a species. A second source of genetic variation is spontaneous change in gene structure; such changes are called mutations. Since genes encode information for the production of the phenotype, recombination and mutation are two sources of phenotypic variation among individuals of a given species. This genetic variation is crucial to the operation of natural selection, for natural selection can work only if variations or differences exist among individuals. If all organisms had the same genotype and the same phenotype, there would be nothing to select from and no potential for change.

A third source of phenotypic variation is environmental differences across the geographical range of a species. The particular environment in which an organism develops can affect its phenotype. The same genotype may produce different phenotypes in response to different environments (e.g., nutritional deprivation in early development may result in stunted adults). Conversely, different genotypes may produce very similar phenotypes in similar environments. Thus the adult phenotype is a product of a given genotype controlling development in a given environment.

Evolution is descent with modification or change through time, and the course of evolution is shaped by natural selection. The results of evolutionary change are correlations between phenotypes and the functional demand of environments. For example, swimming animals tend to be streamlined, with tail or limbs modified to provide propulsive thrust. Fliers have light bones and extensive forelimb-supported flight surfaces. Digging or burrowing animals have powerful limbs, and fast runners have long slender limbs. Such correlations illustrate the phenomenon called adaptation—the "good fit" between organism and environment—and the particular features under discussion are also called adaptations (for a particular function or environment). Scientists have commonly assumed that virtually all aspects of an organism have been shaped by natural selection and adapted for a particular way of life. However, there is renewed controversy among evolutionary biologists over whether natural selection is the only significant factor in determining diversity among organisms. Some genes have been shown to be tightly linked to other genes, and changes in one gene can, indirectly, affect these other genes as well. Such linkages can produce phenotypic features that are passive by-products of selection for other features. Second, some phenotypic features may be neutral or insignificant with respect to natural selection. These features may be carryovers from past selection or part of the phylogenetic heritage of an organism. Third, some features may be the result of biomechanical constraints that developed with an increase or decrease in body size.

Evolutionary Phenomena

We can be fairly sure that a feature is an adaptation if it has occurred independently several times in different groups of organisms that are subject to similar selective pressures. This phenomenon of multiple occurrence is called convergent evolution, and examples are the streamlined bodies of swimmers, the wings of flying vertebrates, and the long, slender legs of runners. All are examples of the form-function correlation we discussed in chapter 1. Making this correlation is one approach to understanding the functional significance of a given feature. Convergent evolution, or widespread form-function correlations, provides good, albeit circumstantial, evidence that given features of an organism are adaptations. Corroborative evidence that a feature is an adaptation could be provided by a biomechanical analysis showing that the feature functions effectively in a manner appropriate to its presumed biological role.

When we look closely at convergent features, we often find that the features differ on a fine scale. The flight surfaces of bird wings are composed of feathers, while these same surfaces in bats and pterosaurs (flying reptiles) are skin. The wings of bats are braced by digits (fingers), while the braces of pterosaur flight surfaces were thin slips of muscle. The propulsive tail fins of ichthyosaurs (aquatic reptiles) were oriented vertically for effective side-to-side movements, like tail fins of fishes, while the tails of cetaceans such as dolphins and whales have horizontal planes for up-and-down propulsive movements.

Such differences as these are called alternative solutions to similar problems, and we may see them even where there are no convergent similarities. Antelopes and kangaroos both graze on open grasslands and escape from predators by covering long distances swiftly. But kangaroos hop on hindlimbs that are constructed very differently from those of swiftly running antelope. Alternative solutions come about by chance or historical accident or because natural selection works on what is available in the traits that shape an organism's basic morphology. Morphology differs from group to group, and different organisms have sometimes evolved quite different structural solutions to their common problems of making a living.

When we look closely at virtually any organism, we see a mosaic of primitive and advanced features. The most primitive living mammals are the insectivorans. Yet among these mammals we see specialized forelimbs for digging (moles), specialized incisor teeth for killing insects (shrews), and specialized hair for protection (hedgehog spines). This phenomenon is called mosaic evolution and results because different parts of an animal's body can evolve separately and at different rates, so organisms can be (and are) combinations of primitive and advanced features. Because of the phenomenon of mosaic evolution, any ordering of organisms into a "scale of nature" or general hierarchy of primitive to advanced is bound to be a subjective exercise, and any particular sequence will be dependent on the particular characters that are being emphasized.

The fossil record also demonstrates that rates of evolution, both within and among groups of organisms, vary widely. Some groups appear to have changed very little since their first appearance. The opossum, family Didelphidae, is one such example. There has been little morphological change in animals of this group over the past 70 million years. Other groups appear to have changed relatively rapidly. The cattle and antelope family Bovidae has radiated into a great diversity of species during the past 10 million years. Some groups, such as the jawed fishes, underwent large evolutionary radiations right after the appearance of a major morphological innovation, jaws. Other groups, including the mammals, remained relatively unchanged for a long period of time after their first appearance despite a host of new adaptive features. During the past decade, some paleontologists have debated whether the dominant mode of evolution is one of gradual, steady change or one consisting of long periods of no change (stasis) punctuated by short periods of rapid change. This latter view is called the punctuated equilibrium mode. Unfortunately, there are few groups for which we can measure evolutionary rates over a fine enough time scale to find evidence relevant to this question. Further, the fossil record preserves only a small part of the phenotype, the animal's hard parts, so we cannot know whether other (unpreserved) parts of the body were changing during times of apparent stasis. Thus it is likely to be difficult to use the fossil record to test whether evolution is a punctuated or gradual process.

Scientists have long recognized that the fossil record shows major extinction events: one occurred 250 million years ago, at the Permo-Triassic boundary, and another 65 to 70 million years ago, at the end of the Cretaceous. Now, thanks to two exciting developments of the past few years we have evidence that these mass extinctions may have been caused by the impact of large extraterrestrial bodies, like asteroids or comets, and that mass extinctions may occur with a regular periodicity. The main evidence for the asteroid impact hypothesis has been the discovery of relatively high levels of the trace element iridium, an element normally found only in extraterrestrial bodies. The presence of iridium on Earth in sediments deposited at times of mass extinctions supports the idea that there is a relationship between the impact on Earth of a comet or asteroid and the occurrence of mass extinctions.

The recognition that mass extinction events may occur with a periodicity of 26 or 30 million years has come from a statistical analysis of recently compiled records of ranges of fossil marine organisms. Both hypotheses are still controversial, although corroborative evidence is coming in from independent sources. Now astronomers are searching for a phenomenon that might account for the apparent 26- or 30-million-year periodicity, and biologists are looking for reasons why some groups became extinct during mass extinctions while other groups survived. Paleontologists find the extinction controversey exciting because we now have a totally new hypothesis to account for major faunal turnovers—why, after millions of years of apparent success, dominant groups became extinct and were replaced by new groups. Previously, it was commonly believed that most extinctions resulted from competition by newly evolved or recently immigrated "superior" groups of organisms. Now we must consider the possibility that mass extinctions opened up new ecological niches and that these new opportunities, not competition, allowed the evolution of new groups of animals. For example, mammals existed as relatively minor elements of reptile-dominated faunas for over 100 million years, until the extinction of many groups of reptiles at the Tertiary-Cretaceous boundary. Only then, around 65 to 70 million years ago, did mammals diversify extensively and radiate into many niches previously occupied by reptiles. No key innovations appeared in mammals at that time; rather, for the first time, they had the opportunity to fill many unoccupied ecological niches.

Now, with the basic elements of evolution in mind, we can turn more specifically to the story of vertebrate life and the evolution of vertebrate design.

The Basic Vertebrate Body Plan

The earliest vertebrates lacked hard parts, so our knowledge of their structure comes not from the fossil record but from the study of a composite of features seen in some of the chordate relatives of vertebrates and in the early developmental stages of modern vertebrates. The earliest vertebrates were small, mobile, aquatic filter-feeders, and figure 3.1 shows a schematic plan of what we infer to have been their basic body organization.

The largest organ in the body was the pharynx, a barrel-like structure at the head end that served as a strainer for food gathering. The pharynx had a small opening in front (the mouth), through which the animal sucked water carrying mud and organic debris, and a long row of vertical slits along each side through which the water was expelled. The mud and organic material passed posteriorly into a thin tube, the gut, for digestion and absorption of nutrients. Suction to operate the pharynx was created through the action of muscles in the walls between the slits, which could expand and contract the volume of the pharyngeal cavity, alternately sucking water in through the mouth and squeezing it out the side slits.

The ancestors of vertebrates were small, thin-skinned animals only a few inches long. Because of their small size, they had a high ratio of surface area to body volume. Therefore they could take in enough oxygen and lose enough carbon dioxide waste (respire) through blood vessels in the skin. However, when later vertebrates evolved to larger sizes and/or developed bony scales or plates in the skin, a more complex respiratory system was necessary. This respiratory system was developed through the evolution of sheets of small, tightly folded blood vessels attached to the walls of the pharynx between the slits. These tissue sheets of blood vessels, the gills, increased the size of the surface area available for gas exchange in the path of water flowing out of the pharynx. The muscles and supporting structures at the bases of the gills, in

21

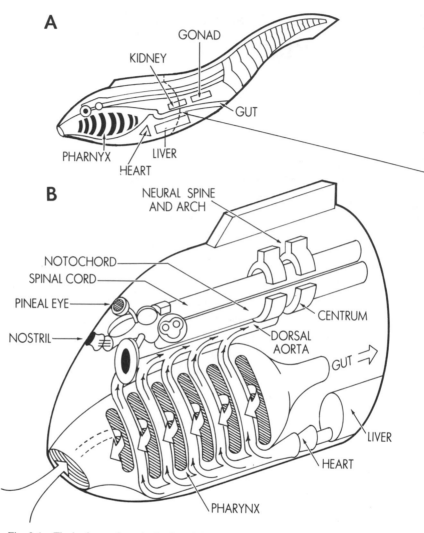

Fig. 3.1   The basic vertebrate body plan. *A*, reconstruction of the ancestral vertebrate. *B*, major systems in head and trunk regions. *C*, top view of body showing how lateral undulations produce backward and sideway thrusts against the water.

the walls of the pharynx between the slits, are called gill arches, and the slits, of course, are the gill slits.

Oxygen and nutrients were carried to all parts of the body dissolved in a fluid called blood, which flowed through a network of muscular tubing called the circulatory system. One main trunk, the dorsal aorta,

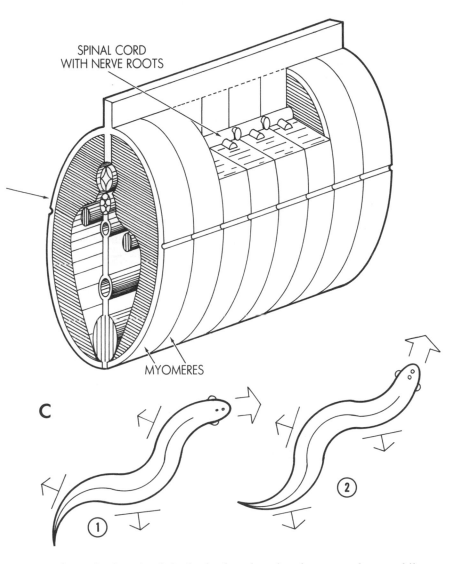

SPINAL CORD
WITH NERVE ROOTS

MYOMERES

C

①

②

ran down the length of the body dorsal to the pharynx and gut, while
another main vessel extended along the body ventral to the gut. Just
posterior to the pharynx, the walls of this lower vessel were greatly
thickened to form a powerful, chambered pump, the heart, that squirted
the blood anterior and dorsal through the gills, where it was oxygenated. The heart was not only powerful, it was also extremely durable,
because it was composed of a special kind of muscle (cardiac muscle)

that is capable of rapidly contracting and relaxing continuously throughout an animal's lifetime. Vessels that carried blood from the heart to the
tissues of the body are called arteries, while those that carried blood
back to the heart (usually with carbon dioxide and other waste products)
are called veins. Exchange of gases and other dissolved materials between the blood and the tissues of the body took place through extremely small, thin-walled vessels called capillaries, which extended
from the ends of arteries to the beginning of veins.

Metabolic waste products other than carbon dioxide were removed
from the blood by a pair of filters, called kidneys, located dorsal to the
gut, on either side of the dorsal aorta. The kidneys also functioned to
maintain the proper salt concentration in the blood, through their capac-
ity to remove excess salts or excess water. Just posterior to the kidneys
and closely associated with them were a pair of organs, the gonads, that
produced, stored, and maintained the reproductive (sperm and egg)
cells.

Food was digested in the gut, broken down by enzymes secreted by
cells in the gut walls. Nutrients were absorbed into the bloodstream
through the walls of the gut and carried to a very large, spongy organ
located on the lower main blood vessel, just posterior to the heart. This
large organ, the liver, converted food molecules to easily stored forms
and also served as a site for food storage. The liver also removed toxic
products and worn-out red blood cells from the bloodstream.

The propulsive system of the earliest vertebrates consisted of a long
series of blocks of muscles, called myomeres, that extended down each
side of the body posterior to the pharynx. The muscle fibers in each
segment were relatively short and ran from front to back, attached to the
tough connective tissue partitions that separated one myomere from the
next. In addition, there was a vertical sheet of connective tissue that ran
down the midline, dividing the myomeres into left and right sides. For
locomotion, the myomeres contracted in a series of waves that passed
from front to back of the animal and alternated from side to side, throw-
ing the body into a series of lateral curves or undulations. Because these
undulations passed backward and, because the body wall pressed backward and sideways against the water (fig. 3.1), the motion gave the
body a forward propulsive force or thrust. (The sideways components of
thrust are equally distributed on the left and right sides and cancel each
other out.) A low fin that extended along the dorsal (top) midline of the
body and around the tail provided some stability for the body during

swimming as well as an increased surface area that could push against the water.

The only major internal supporting structure of the body was a rod of elastic tissue, the notochord, which extended posterior, down the length of the body above or dorsal to the pharynx and gut. Because of its stiffness, the notochord prevented the body from telescoping when the myomeres contracted. Also, because of its elasticity, the notochord acted as an antagonist to the myomeres. That is, it helped straighten out the body after the myomeres that had just contracted and produced a curve in a given section were relaxed.

To synchronize the waves of myomere contraction, there was a major nerve trunk, the spinal cord, that extended posterior down the length of the body just dorsal to the notochord and between the left and right muscle blocks. The spinal cord was hollow, a reflection of the way that it formed in the animal's embryonic stage, when a sheet of nervous tissue folded in on itself to make a tube. From the spinal cord a pair of small nerve trunks ran to each myomere, creating an outgoing pathway that carried motor impulses to activate the myomeres, and an incoming track that carried sensory information such as perceptions of temperature, pain, and pressure to the spinal cord. To protect the spinal cord from damage and to provide a strong site for the attachment of myomere partitions, a series of cartilaginous arches, called neural arches, roofed the spinal cord, one arch per pair of myomeres. Cartilage is the tough, rubbery material that can be felt in your ears and nose tip. Early in vertebrate history a second series of cartilaginous plates, or inverted arches, developed around the notochord below the neural arches. This latter series of cartilaginous elements, called centra (singular is centrum), presumably strengthened the notochord so that it provided better support for attachment of the myomere partitions. A neural arch plus a centrum is called a vertebra (plural is vertebrae), and the entire series of vertebrae is called the vertebral column, or backbone.

At the anterior end of the body, the spinal cord was expanded into a great nervous center, the brain, that received an integrated sensory input from the body and from three special sensory systems that were clustered on the head. The first, the olfactory system (sense of smell), was a detection system that identified things by the size and shape of their constituent molecules. Olfactory receptor cells lined a small sac, the nasal capsule, into which water flowed through a single median nostril at the front of the head. Information from the olfactory receptors passed to paired olfactory bulbs at the front end of the anterior portion of the brain (the forebrain). The second, the visual system, consisted of light-sensitive cells arranged in a sheet (the retina) located in two large lateral eyes and one very small median eye (the pineal eye) on top of the head. In later vertebrates, the pineal eye usually lost its visual function and became mostly glandular. Visual input was received by a paired structure called the optic tectum, located in the roof of the middle portion of the brain (the midbrain).

The third special sensory system was the acoustico-vestibular system, which consisted of two major components (acoustic and vestibular), both of which used similar receptor cells. These receptor cells, called hair cells, signaled any bending of their hairlike processes. The hair cells in the acoustic component were specialized to detect sound waves or vibrations in the water. One set of these sensors was located in a network of grooves on the head and down the sides of the body, in a structure called the lateral line system, that signaled the reception of vibrations from nearby sources. A second set of acoustic sensors was clustered in a pair of capsules, called the otic capsules, located on either side of the back portion of the brain. This latter set of receptors was specialized to detect sound waves from more distant sources, which passed through the tissues of the head. The vestibular part of the acoustico-vestibular system provided the animal with the sense of equilibrium, and it consisted of two sets of hair cells that were also located in the inner ear in the otic capsules. One set of hair cells was arranged to detect turning movements of the head or changes in acceleration from the inertial movement of fluid in long curved tubes called the semicircular canals. The other set was designed to signal changes in position relative to the pull of gravity (i.e., with respect to up or down) as indicated by shifts in the position of small weights located on top of the receptor cells. Acoustico-vestibular signal input traveled to paired centers in the side walls of the back end of the brain (the hindbrain).

In addition to the paired centers in the brain for the three special sensory systems—the olfactory bulbs in the forebrain, the optic tectum in the midbrain, and the acoustico-vestibular centers in the hindbrain—there were also brain centers that received and integrated information from many if not all sensory modalities. These were the cerebral hemispheres, paired outgrowths from the roof of the forebrain, and the cerebellum, a median center in the roof of the hindbrain that was especially important for motor coordination. Like the spinal cord, the brain was also protected by a capsule of cartilage, called the braincase, that formed from several pieces growing together. It also is likely that there were cartilaginous walls around the nasal and otic capsules.

The brain and spinal cord are referred to as the central nervous system, and the paired nerves that extend out from them are the peripheral nerves. Brain and spinal cord plus the peripheral nerves constitute the nervous sytem. Within the nervous system there were two separate parallel sets of wiring, each with its own sensory input and motor output pathways: one for the pharynx, gut, and circulatory vessels, called the visceral system, and the other for the rest of the body, called the somatic system.

The nervous system integrated and regulated body activities, with signals (nerve impulses) that travel its circuits extremely rapidly, in milliseconds. There was, in addition, a second system for the integration and regulation of body functions, with signals that traveled slower and acted over longer time periods. This was the endocrine system, which operated via hormones, chemical signals secreted by glands and circulated throughout the body via the bloodstream. The endocrine glands important in early vertebrates were the pituitary, thyroid, adrenals, and gonadal tissue. The pituitary, located ventral to the middle of the brain, is unusual in that it is a composite gland, with part derived from a ventral outgrowth from the floor of the forebrain and part from a dorsal ingrowth from the roof of the mouth. It secretes hormones that regulate body growth, gonad function, and the activity of other glands.

The thyroid, located originally in the floor of the pharynx, is involved in the regulation of basic body metabolism. The adrenals were originally in scattered patches along the roof of the body cavity; in later vertebrates, they came together as a pair of discrete bodies, one next to each kidney. They secrete adrenaline, a hormone that stimulates circulatory and respiratory systems for "fight or flight" situations (when danger threatens). Glandular tissue in the gonads secretes hormones that regulate gamete (egg and sperm) development and affect other parts of the body that may be involved in reproductive behavior or function.

There were two more systems of the body, one so obvious it is often overlooked and the other not obvious at all. The first and obvious is the skin, or integument, which is what one sees at first glance since it is the covering of the vertebrate body. The skin is composed of two layers, an outer epidermis and an inner dermis. These layers are formed from different embryonic tissues, the epidermis from what is called ectoderm and the dermis from mesoderm. Epidermis and dermis interact in embryonic development to produce, in various vertebrates, a great variety of body coverings (e.g., bony plates, horny scales, feathers, or hair) as well as many different kinds of glands (mucus, poison, sweat, scent, or milk glands). The skin serves the obvious function of protecting the body from damage and from invasion by viruses, bacteria, and other parasites. It also helps maintain proper water balance and houses a variety of sensory receptors such as those for pain, touch, and temperature.

The last and not very obvious system of the basic vertebrate body is the immune system, which is not at all apparent because it has no obvious single structure. It is composed of cells called leukocytes, which are specialized to detect and kill viruses and bacteria that have invaded the body. There are many different kinds of leukocytes and they are found in the bloodstream (some are called white blood cells) and in all parts of the body. Leukocytes are attracted to areas of infection or damage, where they seek out and destroy foreign organisms and also help repair damaged tissues. Some leukocytes are specialized to manufacture specific ("custom-designed") proteins, called antibodies, that neutralize specific foreign organisms or their proteins, called antigens. Antibodies can be maintained throughout an individual's life as a biological record of past invasions and as protection against future attacks by the same invaders.

Throughout this description of the basic body plan of the first vertebrates, you may have recognized certain systems and structures as familiar. The reason for their familiarity is that many modern vertebrates, including humans, retain a good deal of this basic plan and many of the same body systems as well.

That completes our description of what we infer to have been the basic body plan and systems of the earliest vertebrates. We have constructed a small, superficially wormlike, soft-bodied animal, probably no more than a few inches long, that wriggled through the mud and sand at the bottom of shallow bodies of water and made a living by filtering the water for food particles. This is a hypothetical reconstruction, although it is based on features seen in some of our primitive chordate relatives and in our own embryos. Now let us turn to the fossil record and see what we can learn about the oldest known vertebrates.

CHAPTER 4

The Earliest Fossil Vertebrates: The Ostracoderms

The earliest vertebrates probably first evolved about 600–650 million years ago, or between November 9 and 12 of our imaginary year of the age of the Earth. This was a time of great evolutionary radiations, and animals with many different basic body plans appeared for the first time in the shallow seas that covered much of the continents. Many of the major groups of modern marine invertebrates, as well as many now-extinct groups, can be traced back to that period. One of the fascinating major questions about the history of life on Earth is why this evolutionary explosion took place at this time, but so far no obvious answers are apparent.

The very earliest vertebrates were soft bodied and as yet have not been found as fossils. The oldest fossil evidence of vertebrates dates back only to about 520 million years ago, or around November 21. The fossil evidence, scraps of bony scales, is enough to indicate the existence of vertebrates at that time, but it provides no direct information on what they were like. It is not until November 29, or around 415 million years ago, that we get our first good fossil evidence of early vertebrates.

At this time, at least 105 and possibly as much as 235 million years after they first appeared, the vertebrates had diversified into at least four major groups. They were mostly relatively small animals, ranging from a few inches to a foot or so in length, and most were covered with heavy bony plates and scales, from which they get their name of ostracoderm, or shell skinned. Underneath the bone, ostracoderms appear to have been constructed along lines similar to the basic vertebrate body plan described in chapter 3. They differed strikingly from most modern fishes, in that they lacked jaws and a regular pattern of paired fins.

One of the most common groups of ostracoderms were the osteostracans (bone shelled). They get their name from the broad, convex roof of solid bone that covered the head end of the body, including the pharynx (see fig. 4.1). This "shield" had openings for two close-set

29

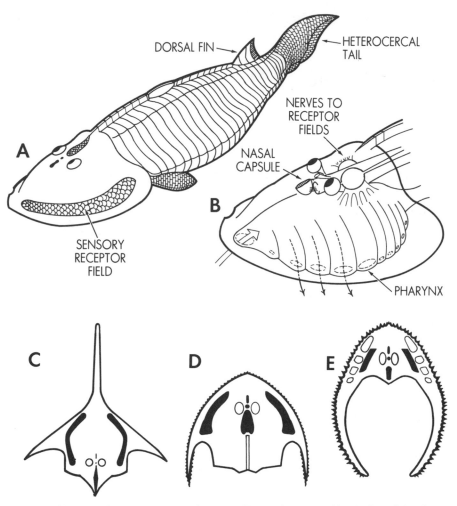

Fig. 4.1 Osteostracan ostracoderms. *A, Hemicyclapsis murchisoni,* about 8 feet long (after Stensio, 1939; Piveteau, 1964). *B,* cutaway of head shield to show brain and pharynx. Diversity of osteostracan head shield shapes: *C, Boreaspis ginsburgi* (after Janvier, 1977). *D, Thystes egertoni* (after Stensio, 1932). *E, Sclerodus pustuliferus* (after Stensio, 1932).

eyes (which looked straight up), plus a small pineal eye between them and a single median nostril in front of the pineal. In addition, there were usually three elongate areas of tiny scales, two lateral and one at the back of the shield. The underside of the body beneath the head shield was covered by a flexible sheet of tiny scales, with a round mouth opening at the front and a series of up to ten pairs of round gill openings along the edges. In some osteostracans, a thin layer of bone was deposited around the brain and the major cranial nerves and blood vessels. This fortunate circumstance preserves details of osteostracans soft anatomy that we would otherwise know nothing about. We can see that there was a single nasal capsule just anterior to the brain (as the single median nostril would suggest) and only two semicircular canals in the otic capsule (part of the vestibular system), in contrast to the three semicircular canals seen in almost all living vertebrates. Further, there were large nerve trunks that connected the three elongate areas of tiny scales to the otic capsules at the back of the brain.

Posterior to the head shield, the body was triangular in cross section and covered by rows of vertical platelike scales. Early osteostracans had relatively long, deep shields that covered much of the trunk and no paired fins. Later species tended to have shorter and flatter shields, and most developed a pair of small, scale-covered, paddle-like structures at the back corners of the shield. In addition, many species had a median dorsal fin toward the rear of the trunk, and a few had a second dorsal fin more anteriorly located. The body terminated in an asymmetrical tail, with a large muscular dorsal lobe, into which the notochord extended, and a small flexible ventral lobe. This type of tail is called heterocercal and is the type seen in the earliest jawed fishes and in modern sharks. The flattened head shield of the osteostracans, with eyes and nostril on top and mouth on the underside, suggests that they were bottom dwellers. This inference comes both from form-function correlation and optimal design analysis. Many modern bottom-dwelling animals such as skates, rays, catfish, and horseshoe crabs show roughly similar body shapes to that of the osteostracans. Optimal design analysis suggests that a broad flattened shield would be the most stable shape for living on the bottom.

We reach this conclusion by looking at the water flowing along the bottom of the sea as though it were water flowing through a pipe. When we do this, one of the basic principles of fluid dynamics (Bernoulli's principle) tells us that, as a fluid moves along in such a system, energy is conserved. That is, at any place along the tube, the kinetic and potential energy must sum to the same value. Kinetic energy is measured by the velocity of the fluid while potential energy is measured, in part, by pressure. In such a system, then, if the velocity of the fluid increases at any point along the pipe, then the pressure must decrease at that point to conserve energy. A second principle of fluid dynamics, the law of continuity, states that in such a system the volumetric flow in must equal the volumetric flow out. That is, the cross-sectional area of the pipe times the velocity of the water must be the same at any point along the pipe. Given these two principles of fluid dynamics, it becomes apparent why the optimal design for a bottom-dwelling organism might be a flattened disk.

Think of any organism living on the substrate. It makes some intrusion or creates a partial blockage in the cross-sectional area of our theoretical pipe. Because the pipe has a smaller cross-sectional area at that point, the velocity of the fluid increases over the body and the pressure over the organism becomes lower. These changes occur because of the law of continuity and Bernoulli's principle. The result of the lower pressure on the top of the animal is that a pressure differential is created between the top and bottom of the organism. The pressure below exceeds that above, and the body experiences lift. An organism that relies on being close to the substrate to obtain food through filter feeding must try to minimize lift; otherwise, it will spend a great deal of muscular energy, in the form of fin movement or body undulation, just to stay close to the bottom. Additionally, most bottom-dwelling fish rely on camouflage as a defense against predators, or they are themselves ambush predators that rely on being inconspicuous to their prey. One important aspect of camouflage as a defense is motionlessness. The optimal design to allow a fish to sit motionless and close to the bottom is a flat shape. The flatter the animal, the less lift is created as the water flows over and around its body, and the less fin movement is needed to keep the animal in position.

Earlier we pointed out that primitive soft-bodied vertebrates swam by lateral undulations, with the body pushing against the water. Large scales on the posterior part of the osteotracan body would limit the fine control of these undulations. The large, heavy head with its rigid shield would have made fine control of swimming difficult (imagine pushing a large wooden box ahead of you as you swim!). From these considerations, we conclude that these fish were neither fast nor agile swimmers. They probably wriggled along in the mud and sand of shallow seas and ancient river bottoms, sucking up organic-rich detritus through their jawless mouths and straining out the water in their large pharynxes. Presumably, expansion and contraction of the pharynx provided the suction and controlled water flow.

The osteostracan head shield was a single piece of bone, with no growth lines, which suggests that these fishes went through an unarmored larval stage and then grew the shield only after they reached adult size. There was a fair diversity of head shield shapes among the osteostracans, with a variety of lateral and sometimes anterior "horns" or processes (fig. 4.1). Perhaps the different shield types were correlated with the requirements of living on or moving through different types of bottom sediments or water currents. We are hampered in trying to solve this question of the significance of shield shapes by the lack of living analogues, and the relevant optimal design experiments (like analyzing the performance of models in flow chambers or on different sediment types) remain to be done.

Another puzzling question about the osteostracans is the function of the three elongate areas of tiny scales on the head shield. One suggestion is that they housed electrical field receptors, a sensory system alien to us but recently discovered to exist in sharks, lampreys, and some other modern fishes. Another suggestion is that they contained vibration detectors, perhaps similar to the lateral line system, which appears to have been poorly developed in osteostracans. In either case, we would expect to find the observed nerve connections to the otic capsules. Another question is how the osteostracans oriented their bodies for respiration when they were not feeding. The head shield must have been tilted up enough to allow a respiratory water current to be sucked into the ventrally located mouth.

A recently discovered group of ostracoderms that appear to be related to the osteostracans are the galeaspids (helmet shields). They had head shields similar to those of the osteostracans but with a different pattern of openings. There were two round holes for the eyes, a large median opening (of variable shape in different species of galeaspids) at the front of the shield, and no trace of a nostril or the three elongate fields of tiny scales that are such a constant feature in the osteostracans. On the underside of the head, there was a long series of round gill openings along the edges, as in the osteostracans. The similarities in body form between galeaspids and osteostracans suggest that they shared a similar way of life as bottom-dwelling, detritus feeders. A major unanswered question about the galeaspids is the function of the large median opening at the front end of their head shields. This opening has an unusual position for a mouth, since in most bottom dwellers that opening is located on the underside of the head shield, and it is too large to have been a nostril.

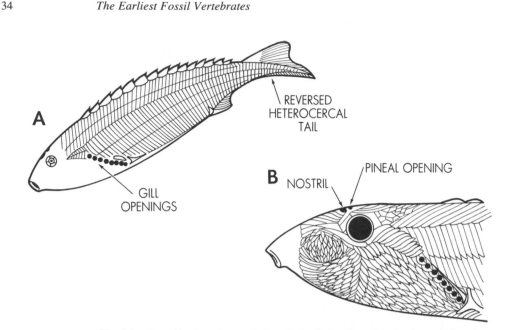

Fig. 4.2   Anaspid ostracoderms. *A, Pterolepis nitidus,* about 4 inches long, with scales omitted from head region (after Piveteau, 1964). *B,* head region of *Birkenia elegans,* with scales omitted from around mouth (after Heintz, 1958).

A second major type of ostracoderm was the anaspids (without shields). Anaspids were small animals, mostly about 4–6 inches long, covered with very small bony scales (fig. 4.2). They had laterally placed eyes and, like the osteostracans, a small pineal eye and single nostril on top of the head. The jawless mouth was a vertical elliptical opening at the front end, and water exited the pharynx from a row of six to fifteen circular gill openings behind the eye. Anaspid bodies were streamlined and flattened from side to side. By streamlined we mean that the body had a torpedo-like shape, rounded anteriorly and tapering gently in the rear. The tail was asymmetrical but unlike that of the osteostracans in that the notochord extended posterior into a larger ventral lobe (a reversed heterocercal tail). Many species had a row of triangular spines along the dorsal midline of the body, and some developed a pair of long lateral skin folds, or fins, along their lower flanks.

The streamlined, laterally flattened body form of the anaspids suggests that they were active swimmers, at least more active than the osteostracans and galeaspids. We can base this conclusion on a long series of studies that have looked at both form-function correlations and at the effects of body shape on swimming efficiency. When a body moves through a fluid, the fluid exerts a force called drag on the body in the opposite direction from which it is moving. Animals whose shapes reduce drag are more efficient because they have to use less energy to travel at a given speed through their fluid environment. But what are the more efficient shapes?

Wind tunnel experiments allow us to create patterns of air flow around different-shaped objects and measure drag. These experiments show that a streamlined, torpedo-shaped body experiences the least drag at moderate speeds. The ideal shape, giving the least drag, for a given velocity is one in which the length is about 4.5 times the maximum diameter. We noted earlier that many groups of aquatic vertebrates have independently evolved this body shape. Dolphins, seals, fish, and extinct aquatic reptiles called icthyosaurs (see chap. 14) have an optimal body shape for minimizing drag while cruising under water.

In addition, the effect of fish body shape on drag has been measured by free-fall experiments. Dead fish are dropped down a large vertical tank of water, and the fall of the fish up to terminal velocity is filmed against a background grid. Knowing the weight of the fish and the velocity of the fish, it is possible to calculate the drag force on the body. These data on drag forces and behavioral observations of living bony fish have allowed us to demonstrate a form-function correlation between low drag coefficients (i.e., streamlined bodies) and relatively high speed swimming. Fish such as tuna, salmon, and trout all have streamlined bodies and are known to be relatively high speed swimmers.

The anaspids had no large rigid head shield to hamper their control of swimming movements, and the diversity of dorsal and ventral fin patterns suggests evolutionary experimentation with stabilization and control. Both of these characteristics support our inference of active swimming. Paleontologists used to think that the reversed heterocercal tail must have produced an upward thrust and that the anaspids were surface feeders. However, studies during the past decade on tail shape and locomotion in modern fishes have revealed that a fish can produce either upward or downward components of thrust regardless of the type of tail: everything depends on the orientation of the flexible fin surfaces attached to the tail. The presence of a heterocercal tail does not by itself indicate surface feeding. The small, jawless mouth of the anaspids suggests that they were filter feeders or that they sucked in small food particles. If they were indeed active swimmers, they probably took in material suspended in the water instead of browsing along the bottom.

Another major group of ostracoderms were the heterostracans (different shelled). The front third to half of the body of these animals was encased in a heavy bony covering, composed of several individual plates that show growth lines (fig. 4.3). There were two small, laterally placed eyes and a very small median pineal eye (not always open to the outside), but there is no trace of a median nostril. Some heterostracans have paired impressions inside the shield, and these may indicate the presence of paired nasal capsules (in contrast to the single median nasal capsule in osteostracans), while other heterostracans are believed to have had just a single median nasal capsule. In either case, water must have entered the nasal capsules through the mouth. The mouth was located on the underside, near the front of the shield, and a row of very small bony plates formed its posterior border. Water passed posteriorly through the gill slits of the large pharynx into a tunnel-like chamber and exited through a single opening on each side at the posterior end of the shield.

The body posterior to the shield of the heterostracans was covered with small scales. Earlier studies suggested that the tail was of the reversed heterocercal type, but recent reconstructions of some species suggest that the tail was symmetrical, with low, flexible fin surfaces above and below a notochord and muscular axis that extended straight back.

Although the trunk was covered with small, light scales, suggesting a relatively flexible posterior body, the fact that most head shields were large and rigid suggests that most heterostracans were bottom dwellers and not good swimmers. Like the osteostracans and galeaspids, they probably wriggled along on or through the mud and sand of ancient sea and river bottoms, sucking up organic debris. The small plates along the posterior edge of the mouth may have supported a mobile, flexible border that allowed some degree of selective feeding or that perhaps could be protruded like a scoop.

There was a great diversity of shield shapes among heterostracans. Some were long and narrow and others short and wide; some had long snouts anterior to the mouth, and others were blunt ended (fig. 4.3). Most were moderately to extremely flattened, like the head shelds of the osteostracans. Some heterostracan shields had no eye openings, indicating that the animals were blind. Others had such tiny mouths that it

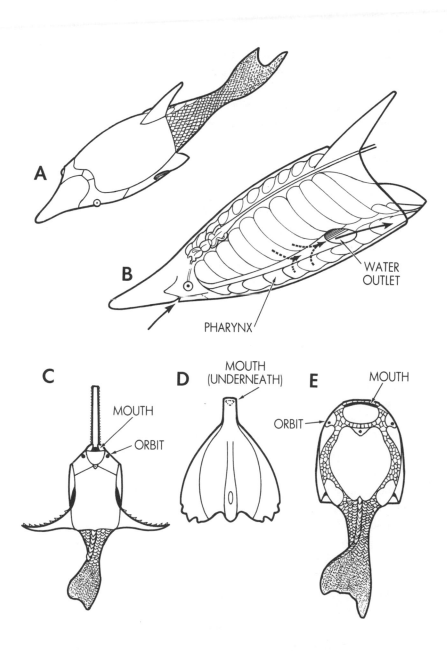

A

B

WATER
OUTLET

PHARYNX

C

MOUTH

ORBIT

D

MOUTH
(UNDERNEATH)

E

MOUTH

ORBIT

Fig. 4.3 Heterostracan ostracoderms. *A, Pteraspis rostrata*, about 9 inches long (after White, 1935). *B*, cutaway of head shield to show brain and pharynx. Diversity of heterostracan head shields: *C, Doryaspis nathorsti*, 5–10 inches long (after Heintz, 1968). *D, Eglonaspis rostrata* (after Obruchev, 1967). *E, Drepanaspis gemuendensis*, over 1 foot long (after Obruchev, 1943).

is difficult to imagine how the animals could have sucked in enough food for body growth and maintenance, although they must have managed somehow.

Some of the heterostracans grew quite large, 3–6 feet in length, which seems like a gigantic size for an animal making a living as a bottom-dwelling detritus feeder. Perhaps some day we will find unusually well preserved fossil specimens that will give us clues to their internal anatomy and possibly some answers to these unanswered questions. The diversity of hornlike structures, some similar to those seen in osteostracans, have been variously interpreted as stabilizers for swimming, anchorage for life in fast-moving streams, devices to stir up mud and food particles, or props to keep the mouth up off the bottom. Optimal design experiments to test these hypotheses have yet to be done.

Heterostracans differ from osteostracans and anaspids in the absence of a median nostril and, perhaps, in having paired nasal capsules. In addition, the microscopic anatomy of the bone making up the shields differs between the two groups (hence the name heterostracan, i.e., different shelled). They did have a single external opening for water from the gill slits. For these reasons, the ostracoderms are usually divided into two major groups, one including all the anaspids and osteostracans and the recently discovered galeaspids and one including all the heterostracans.

The last group of ostracoderms are the coelolepids (hollow scaled). These are small, poorly known forms whose bodies were covered with tiny scales. They appear to have been flattened in front, with a small terminal mouth, but, because the scales are so small, the body outline and head openings are not well preserved and little can be inferred of coelolepid anatomy or their appearance in life. They indicate that there was a group of ostracoderms in which the bony scales and plates had either become very reduced or perhaps never developed very far in the first place. The lack of a heavy bony casing suggests that the coelolepids, like the anaspids, may have been active swimmers.

Thus we see that ostracoderms were a rather diverse group of fishes, and their fossils are fairly common in sedimentary deposits laid down under ancient shallow marine and freshwater seas from around 415 to 375 million years ago (November 29 to December 3). The major advances they show over the presumed ancestral condition (seen in the basic vertebrate body plan) are the development of a sizable tail (or caudal) fin and their covering of usually thick bony plates and/or thinner scales. The larger tail fin increases the amount of surface area the fish can move against the water with lateral body undulations, and this provides increased propulsive thrust. In subsequent chapters, we will repeatedly see the evolution of laterally compressed tails in aquatic animals. This shape maximizes the amount of surface area available to generate propulsive thrust. Furthermore, experiments on modern bony fish indicate that increased caudal fin area and depth correlate with a specific aspect of locomotion, fast-start acceleration. The cruising speed and acceleration of a series of bony fish were compared by increasing the caudal fin area and depth but holding other aspects of body shape constant. Increasing caudal fin area and depth significantly improved fast-start acceleration performance but had little effect on cruising speed. Ostracoderms may not have had a fast-start acceleration comparable with many modern fishes, but the increased caudal fin area may have improved performance relative to more primitive vertebrates with less tail area.

The flexible fin surfaces of the tail allow the addition of upward and downward components to the main forward thrust and thus permit some control of thrust direction. The biomechanical significance of the difference between heterocercal (body axis bent upward and extending into the dorsal tail lobe) and reversed heterocercal (body axis extended ventrally) type of tail is not apparent, since both types can produce both upward and downward thrust. It has been suggested that the heterocercal type would be advantageous for bottom dwellers, since the dorsal muscular lobe would be clear of the sediment. However, we find either reversed heterocercal tails (old interpretation) or symmetrical tails (new interpretation) in the bottom-dwelling heterostracans, and the early members of all jawed fish groups, most of which were free swimming, had heterocercal tails (see chaps. 5 and 6). Perhaps it was chance or historical accident that determined which tail type developed in a particular group of fishes. If there is no biomechanical advantage to one type over the other, we may consider them to be alternative solutions to a similar "problem," that of increasing propulsive thrust.

Why bone evolved is another question. Bone is a tissue unique to vertebrates, a strong, hard material that is a composite of inorganic calcium phosphate mineral crystals plus organic collagen fibers. The fossil ostracoderms indicate that bone first evolved as an external rather than an internal skeleton, and, because it was deposited in the skin, ostracoderm bone is called dermal bone. In many ostracoderms, the outer layer of the dermal bone was composed of a slightly harder but similar material called dentine. In some, there was a very thin outer coating of a still harder material, similar to enamel.

The most popular hypothesis as to why bone evolved is that it provided protection against aquatic invertebrate predators such as the giant "water scorpions" (eurypterids) or nautiloid cephalopods (squid relatives) that coexisted with early ostracoderms. This idea is reflected in the use of terms like "armor" or "shields" to describe the ostracoderm exoskeleton. Another, more recent, hypothesis is that bone evolved to serve as a storage site for calcium and phosphorus, both of which are necessary for muscle activity (and other metabolic processes). Although the heavy dermal bone of most ostracoderms suggests that they were not good swimmers, some may have lived in rapidly flowing streams or areas of strong tidal flow, environments that required them to use constant muscle activity to maintain position, and hence they would need ready access to stores of calcium and phosphorus. The most recent and perhaps most ingenious hypothesis for the origin of bone suggests that it was originally deposited around the network of canals that housed an electroreceptor sensory system, to serve as an insulator and to provide a rigid framework for the network. However, modern fishes with an electroreceptor system lack such an insulating framework. It is often difficult to say why a particular feature first evolved, and certainly the bone of ostracoderms could have served all three of the functions suggested for it thus far. In fact, a feature may evolve to serve more than a single function.

The diet of ostracoderms was limited by their lack of jaws, which probably restricted them to filter feeding or sucking in small food particles. Their heavy bony covering and lack of a regular system of paired and median fins suggest that they were not very efficient swimmers. Nevertheless, they existed for at least 150 million years, and for the last 50 million of this span they coexisted with a great variety of jawed fishes that were better swimmers and had access to a greater variety of food resources. Thus we really cannot think of the ostracoderms as biologically inferior types. Their diversity and long history of life on Earth are good indicators of their success in their particular ways of life.

There are two modern surviving groups of the great ostracoderm radiation, lampreys and hagfish. These are elongate, jawless fishes that lack paired fins, and, unlike the other known ostracoderms, they also lack a well-developed tail fin. They have no bony tissues in or on their bodies. Lampreys have a single, dorsal median nostril, a pineal eye, two semicircular canals, and a long row of circular gill pouch openings (like the osteostracans and anaspids). Hagfish have a single terminal median nostril, a pineal organ that is not exposed at the surface, what appears to be single semicircular canal (probably reduced from an original two), and a variable number of external openings for the gill pouches. The internal skeleton of lampreys consists of a cartilagionous braincase, rudimentary vertebral column and gill arches (the latter a nonjointed, basket-like structure), and a notochord. Hagfish lack vertebrae, and their internal skeleton consists only of the notochord and a partially developed cartilaginous braincase and vestiges of gill supports. Both lampreys and hagfish have the gills located medial (internal) to the gill supports and musculature. Lampreys have a funnel-shaped mouth lined with small horny denticles (projections) that allow them to attach themselves to the bodies of other fishes. They have a muscular, tonguelike structure, also studded with horny denticles, that is used to bore through the body walls of their living hosts. Thus they make a living as parasites that eventually may kill their hosts. Hagfish are burrowers that live mainly on marine worms but also scavenge dead fish. They have sharp-edged horny plates

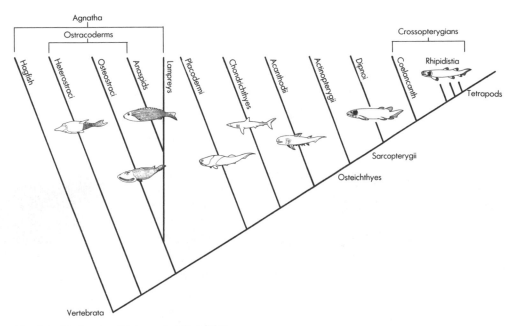

Fig. 4.4   Phylogeny of living and extinct fishes.

set in muscles inside their jawless mouths that function as teeth. A fossil lamprey has been found in deposits almost 350 million years old (December 4). It apparently lacked the rasping mouth apparatus seen in modern lampreys, suggesting that the earliest lampreys were not parasites.

All jawless fishes are usually placed in the class Agnatha (without jaws), which is often divided into two subclasses: the early armored types (ostracoderms) and the modern lamprey and hagfish (called cyclostomes, or round mouths). More recently, scientists who specialize in studying these jawless fishes have considered such a classification as artificial and not an accurate reflection of the phylogenetic relationships among ostracoderms. Current classifications usually recognize an early dichotomy between osteostracans, anaspids, and lampreys on the one hand and heterostracans, coelolepids, and hagfish on the other (fig. 4.4). However, so little is known about the soft anatomy of the ostracoderms that we may expect revisions in these ideas as new information becomes available.

The First Jawed Fishes

Jawed fishes first appear in the fossil record between 410 and 430 million years ago (November 28–30), about the same time that we have good fossil evidence of the jawless ostracoderm radiation. The evolution of teeth and jaws was a tremendous breakthrough for the vertebrates, for it meant access to a much greater variety of food items than was available to the ostracoderms. Also, jaws opened possibilities for defense and manipulation of objects in the environment that were not possible for ostracoderms.

We know from embryological and comparative anatomical evidence that jaws evolved from an anterior pair of gill arches. The embryological evidence for the origin of jaws comes from a comparison of the embryonic tissues that give rise to the parts of the head. Jaws and their associated muscles and gill arches and their associated muscles show a common embryological origin distinct from that of the rest of the head.

The internal supports of both gill arches and jaws are derived from a very special kind of embryonic tissue called neural crest. Neural crest consists of cells that bud off of the developing neural tube (early stage brain and spinal cord) and migrate through the embryo to form not only the jaws and gill arches but also important ganglia (cell clusters) of the spinal cord nerves, pigment-producing cells for the skin, and parts of the braincase. The muscles of both the jaws and the gill arches are derived from lateral plate mesoderm (an unsegmented sheet of embryonic tissue). The rest of the braincase and all other parts of the internal skeleton of vertebrates as well as the rest of the body musculature are formed from segmented blocks of embryonic mesoderm called somites. The common embryonic origin of jaw and gill arch skeleton from neural crest cells and of jaw and gill arch musculature from lateral plate mesoderm is strong evidence that jaws were once gill arches.

The location of the few pieces of braincase that form from neural crest cells and the pattern of major nerve trunks in the cranial region

43

have suggested to some embryologists that the original first gill arch became incorporated into the braincase when jaws evolved. This would mean that the jaws represent the greatly transformed original second gill arch.

The comparative anatomical evidence that jaws were derived from gill arches comes from examination of these structures in modern sharks. Sharks retain a relatively primitive pattern of basic head organization, in which the skeletal elements, nerves, and muscles of the jaws appear to be a serial continuation of what is seen in the gill arches (fig. 5.1). Also, between the jaws and the first gill arch behind them (called the hyoid arch), sharks and primitive bony fishes retain a small round opening, the spiracle, that appears to be a remnant of what was once a complete gill slit.

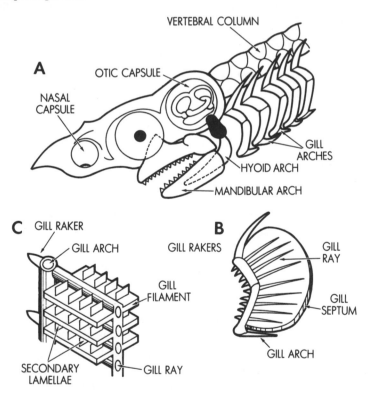

Fig. 5.1   From gill arch to jaws. *A*, modern shark, showing skeleton of head and pharynx; the hyomandibula is colored black. *B*, shark gill structure. *C*, enlargement of gill septum.

The evidence that jaws evolved from gill arches is convincing, but there is an interesting problem when we try to envision what gill arches were like in the ancestral jawless vertebrates. In the living jawless fishes (lampreys and hagfish), the gill arches are external to the gills and the supporting skeleton is an unsegmented, continuous, basket-like structure (reduced in the hagfish). In jawed fishes, the gill arches are internal to the gills, and the supporting skeleton is segmented rather than a continuous network (fig. 5.1). Given these differences in the skeletal system of the gill apparatus, it has seemed extremely unlikely that either of these two types of gill construction could have been derived from the other; rather, they have been thought to represent two alternative solutions to the problem of developing a gill apparatus, and it has seemed that the common ancestor must have been far back in vertebrate history, perhaps before gills evolved. However, recently, researchers have pointed out that the details of gill anatomy are extremely similar between the living jawless and jawed fishes, so similar that it seems most likely that a basic vertebrate gill was present in the common ancestor and that it was only the arches, or skeletal supports, that evolved independently. It would be extremely interesting to know how the gill arches were constructed in the ostracoderms and critical for any hypotheses of phylogenetic relationship to the jawed fishes, but, unfortunately, so far the fossil record has not yielded that information.

When an anterior gill arch became enlarged to form the jaws, the upper portion of the gill arch behind it (the hyoid arch) also was enlarged, and in the earliest jawed fishes that segment of the hyoid arch (called the hyomandibula) forms a brace between the back of the jaws and the braincase. The lower portion of the hyoid arch helps to support the floor of the mouth and in later vertebrates provides anchorage for the tongue. In some groups, particularly those that specialized on relatively hard food items, the upper jaws became fused to the braincase for firmer support, and the hyomandibula, no longer functioning as a brace or strut, became reduced in size. In other groups, the hyomandibula remained a large element, linking the jaws to the braincase.

The development of jaws allowed a dietary shift away from filter feeding, and the pharynx no longer functioned as a strainer. It became shorter, with a reduced number of gill arches that were compactly arranged in an oblique, nested arrangement. The development of large, strong jaws also made it possible to take in large food items, and it was presumably in tandem with the evolution of jaws that an anterior portion of the gut became enlarged into an expandable, tough sac, the stomach, which can hold large chunks of food while they are being digested, and which has a lining able to withstand the corrosive effects of its very strong digestive enzymes.

Concurrent with the evolution of jaws was the development of teeth. In their simplest form, teeth were (and are) hard, conical structures that line the jaws. They are composed of dentine with a thin outer covering of a hard enamel-like substance and a base of bone. Their composition and mode of development (via epidermal and dermal interactions) suggest that teeth are modified dermal scales. Even in early jawed fishes, teeth had a variety of shapes, from simple narrow cones, optimal for puncturing and holding, to bladelike structures that enhance shearing and cutting, to broad, flat crushing surfaces. In most groups of fishes, worn teeth were replaced continuously throughout life.

There have been four major groups of jawed fishes, the now-extinct acanthodians and placoderms, and the still-thriving cartilaginous and bony fishes (fig. 4.4). In addition to jaws, they all have paired nostrils and nasal capsules and three semicircular canals in the vestibular system. Their myomeres were divided into upper and lower portions by a horizontal partition, they have some system of paired fins and the earliest members of each group had a heterocercal tail.

Ostracoderms, as discussed in chapter 4, show a fair diversity of fin and finlike shapes and positions. Similarly, the jawed fishes have evolved a number of different tail shapes, types of dorsal and anal median fins, and paired fin positions. An interesting unsolved question is the relationship between specific patterns of fin placement and morphology and the characteristics of fish locomotion. At this point all we know is that, generally, median and paired fins and different caudal fin shapes are related to problems of stability, speed, and maneuverability during locomotion.

Fish are made up of bone and muscle, materials that are denser than water. Without some compensating mechanism a fish body will sink in the water column. We refer to this as being negatively buoyant. The early, extinct fishes were probably all negatively buoyant, and so are many of the living fishes. This is one source of instability.

The gravitational forces acting on any object can be summarized by a single force acting through a point called the center of mass, the position of which depends on the distribution of weight in the body. Similarly we can identify a center of buoyancy for an object. If the object is made of a single material of uniform density, then its center of mass and its center of buoyancy will be the same. If the body is not made up of components of a uniform density (think of the head shield versus the muscular body parts of an ostracoderm for an extreme example), the center of buoyancy will be located at some distance from the center of mass. The pull of gravity acts through the center of mass and is a downward vertical force. Buoyancy, due to water displacement by the object, is an upward vertical force acting through the center of buoyancy. The result of these forces acting at two different points of the body is to produce a torque or turning moment on the body, if the forces are unbalanced. The body will have a tendency to turn in the direction of the higher force. Instability in which the body rotates around its own transverse axis is referred to as pitching. So, even without considering locomotion, we have described two sources of instability for a fish: that related to its negative buoyancy and that related to the differential density of its body. It seems that one of the functions of the median and paired fins is to counter these two types of instability, by producing forces to oppose the force of gravity and the torques arising from the separation of the centers of mass and buoyancy.

Next consider what happens when a fish swims. If the propulsive force acts through the center of mass of the animal in a horizontal direction, and the weight of the body is being pulled downward in a vertical direction due to gravity, the resultant force moves the fish forward and down in the water. The fish loses its horizontal position. Similarly, if the thrust of the propulsive fin(s) does not pass through the center of mass, a torque will be produced, rotating the fish in the horizontal or vertical plane. One way to counter these movements is to provide lift (or an upward vertical force) at the same time that the propulsive, horizontal thrust of the caudal fin is generated. Paired pectoral fins can do this. Water passing over the profile of the pectoral fin changes velocity for the reasons discussed in chapter 4 (Bernoulli's principle and the law of continuity). This leads to a pressure differential above and below the pectoral fin producing a lift force.

Another way to compensate for horizontal instability is to have the direction of the propulsive force from the tail act at an appropriate positive angle relative to the horizontal. Different directions of propulsive thrust can be generated by asymmetrical caudal fin shapes. As we look at the different fin patterns of various fish groups in this and the following chapters, it will become apparent that a wide diversity of mechanisms have been evolved to deal with problems of instability in fish locomotion.

Acanthodians

The first jawed fishes to appear in the fossil record are the acanthodians (spiny ones), known from 430 to 260 million years ago (or from November 28 to December 11). Most acanthodians were small, from 4 to 6 inches long, with streamlined, laterally flattened bodies (fig. 5.2). Most were covered with very small, diamond-shaped, nonoverlapping dermal bone scales, which were enlarged and formed small plates in the head region, while some species were scaleless. The eyes were relatively large and forwardly placed, leaving little room anteriorly for the nasal capsules, which must have been relatively small. The jaws were relatively long and usually were lined with small, sharp teeth. There

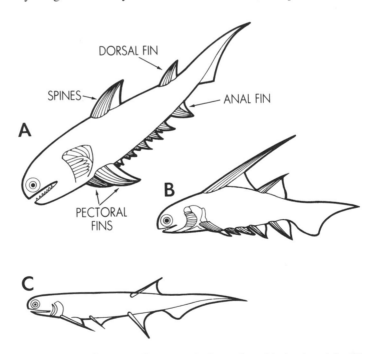

Fig. 5.2   Acanthodian fishes. *A, Climatius reticulatus,* about 6 inches long (after Watson, 1937). *B, Parexus incurvus,* about 4.5 inches long (after Watson, 1937). *C, Homalcanthus concinnus,* about 3.5 inches long (after Miles, 1966).

were five gill arches behind the jaws, each with a small flap to cover the gill slit. In addition, some species appear to have developed a large single flap, called an operculum, that was attached to the back of the hyoid arch and extended back to cover most or all of the gill slits. In a few species, the internal skeleton of the jaws and gill arches was ossified or composed of bone rather than cartilage, and the braincase and vertebral column appear to have been partially ossified.

The fins were triangular webs of skin supported in front by a stout, bony spine. The primitive pattern consisted of two median dorsal fins and one median ventral fin (the anal fin, near the back of the trunk), one pair of large fins behind the gills (pectoral fins), and two rows of five or six smaller paired fins along the trunk between the pectoral and anal fins. In later species of acanthodians, there was a tendency to reduce the number of small paired ventral fins and to develop relatively longer and more slender fin spines.

From the streamlined body form of the acanthodians, we can infer that they were active swimmers. This inference comes both from optimal design analysis (no heavy dermal armor, a body streamlined to minimize turbulance) and by analogy with similarly built modern fishes (form-function correlation). This interpretation is further supported by the presence of three rows of fins (one dorsal and two ventral) in the acanthodians, which would have acted as stabilizers and allowed improved control of swimming movements. The long jaws and small sharp teeth suggest that most acanthodians fed on very small fishes or invertebrates. Their large eyes and small nasal capsules indicate that in these fishes vision was relatively important and olfaction relatively unimportant. There was not much variation in the basic acanthodian body plan and inferred way of life except that some of the later species secondarily lost their teeth, had a lengthened pharyngeal region, and presumably reverted to filter feeding on suspended small food items. The large spines seem in some of the later acanthodians (like *Parexus* in fig. 5.2) may have evolved to give protection against being swallowed by larger fishes.

Placoderms

Twenty million years after the acanthodians first appeared in the fossil record, we find evidence of other types of jawed fishes. One group, the placoderms (plate skinned), is known from about 410 to 350 million years ago (or from November 30 to December 4). Most placoderms were medium-sized animals, 1 or 2 feet in length. Heavy plates of dermal bone cover the anterior third to half of the body of early placoderms, which had a joint between rigid head and shoulder (thoracic) portions (fig. 5.3). The remainder of the body was covered with small bony scales or was without dermal armor. The head shield was relatively broad, and it covered the pharyngeal region as well as the

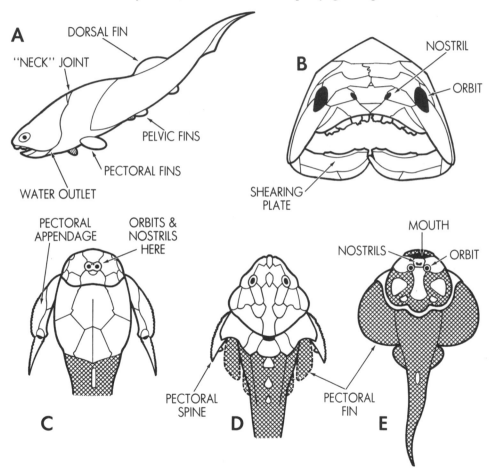

Fig. 5.3   Placoderm fishes. *A, Coccosteus cuspidatus,* an arthrodire placoderm, about 11 inches long (after Miles and Westoll, 1968). *B,* close-up of head armor of *Coccosteus,* to show tooth plates (after Miles and Westoll, 1968). *C, Bothriolepis canadensis,* a bottom-dwelling antiarch placoderm, in dorsal view; note long, jointed appendages (after Stensio, 1963). Progressively more flattened bottom-dwelling placoderms: *D, Lunaspis heroldi;* (after Piveteau, 1969) and *E, Gemuendina stuertzi* (after Piveteau, 1969).

braincase. Water from the gill slits was expelled through a single open-ing on each side of the body between head and thoracic shields. Most placoderms had biting structures in the form of sharp-edged plates of dermal bone lining their jaws, rather than the small pointed teeth of the acanthodians. A stout, spinelike projection extended on each side from the armor in the shoulder region, and there was a pair of fins in the shoulder region (pectoral fins), which attached just behind the shoulder spines. In addition, placoderms had one or two median dorsal fins and a variety of types of posterior ventral fins. The braincase was ossified, and some placoderms preserve evidence of a vertebral column com-posed of ossified neural arches above the spinal cord and ringlike centra elements below the notochord.

Placoderms evolved into a great diversity of body forms in a rela-tively short time. One of the best-known groups, the arthrodires (jointed necks), elaborated the primitive placoderm neck joint by enlarging the gap between head and shoulder shields and developing a pair of ball-and-socket joints at the sides that allowed the head portion to rotate back on the body. Their shoulder armor was shortened, and the slots for the bases of the pectoral fins were elongated. The shearing plates along the border of the mouth tended to become elongate blades, some with sharp piercing points. Most of the arthrodires were a few feet in length, al-though the largest of them may have been close to 20 feet long!

The shearing plates on their jaws indicate that the arthrodires were predators with the ability to attack and kill animals even larger than themselves. The ability to rotate the head shield back on the body could have been used to increase jaw gape for predation on larger organisms (both upper and lower jaws could open simultaneously) and/or to create suction to take in prey or improve respiration by rapidly increasing the volume of the mouth chamber. The enlarged slots for the pectoral fins suggest increased fin mobility. Despite the heavy dermal bone in head and thoracic shields, a good part of the body was unarmored and there-fore flexible enough for unhampered lateral undulations, which sug-gests that at least some of the arthrodires were active swimmers. Indeed, to be effective predators they would need to swim swiftly and with good control.

Another group of placoderms, called antiarchs (fig. 5.3), had dor-sally convex, ventrally flattened head shields; long, boxlike thoracic shields; and limited neck joint mobility. They had dorsally placed eyes and nostrils, and a small ventrally located mouth with small tooth plates. One of their most peculiar features was a pair of slender, armored, jointed appendages—apparently modified pectoral fin spines—that extended laterally from the shoulder region. Paired saclike impressions in the floor of the thoracic shield have been interpreted as evidence of accessory respiratory organs and, if so, would be our earliest evidence for the presence of lungs.

The ventrally flattened head shield, with eyes and nostrils on top and mouth below indicates that the antiarchs were bottom dwellers. The extensive heavy bony covering indicates that they were not good swimmers; they probably wriggled along on the bottom, perhaps using the jointed pectoral appendages to help push or pull themselves along. The small mouth and small mouth plates suggest that antiarchs were detritus eaters, perhaps with a way of life not very different from that of the jawless, bottom-dwelling heterostracan ostracoderms. The small mouth plates would have allowed some degree of selective feeding. If the paired impressions in the floor of the antiarch thoracic shields really do reflect the presence of lungs, we may infer that antiarchs inhabited shallow bodies of frequently stagnant water, where it would have been important to be able to supplement gill respiration by gulping air. This is the habitat where we find modern lungfish who also have functional lungs.

Other groups of placoderms, known as ptychtodonts, petalichtyids, phyllolepids and rhenanids (names as bizarre as their appearances), modified the primitive placoderm body plan by greatly reducing the dermal plates in the head and shoulder regions and by developing variously flattened heads and bodies with dorsal eyes, ventral mouths, and expanded pectoral fins (fig. 5.3). Further, they modified the originally sharp shearing mouth plates to form flat, crushing structures. From form-function correlations with modern fish such as chimaeras (ratfishes), skates, and rays, we can infer that these types of placoderms were bottom-dwelling forms that fed on shelled invertebrates.

Early Bony and
Cartilaginous Fishes

CHAPTER

6

Osteichthyans

Another group of jawed fishes that appeared around 410 million years ago (about November 30), along with the placoderms, were the osteichthyans (bony fishes). The earliest osteichthyans (fig. 6.1) were mostly about a foot or two in length, with streamlined, laterally compressed bodies that were covered with small, thick, bony scales. The scales were enlarged in the head region to form a complex pattern of interlocking plates over the braincase and jaws. Osteichthyans had an operculum or flap over the gill slits, composed of several enlarged dermal plates and attached to the first (hyoid) gill arch. Judging from their modern survivors, the early osteichthyans had a pair of accessory respiratory structures, the lungs. The internal skeleton was relatively well ossified, with bone replacing much of the originally cartilaginous braincase, gill supports, and vertebral elements. The latter included neural arches and neural spines above the spinal cord and inverted arches and spines, called haemal arches and spines, below the notochord around the dorsal aorta artery and a major vein, posterior to the gut. In some groups, there were small centrum elements around the lower part of the notochord.

There was a regular pattern of fins: two sets of paired fins ventrally (larger pectorals anteriorly and smaller pelvics posteriorly), as well as one or two median dorsal fins above and one median anal fin below. The median fins were located relatively posteriorly, close to the tail, and all fins were supported internally by long, thin, flexible rods called fin rays. The paired fins were connected to the body by a U- or V-shaped series of bones, the pectoral and pelvic girdles, which were imbedded in the musculature of the body wall. The pectoral girdle was quite large and included a series of dermal plates that extended up to the back of the skull. The tail was of the usual heterocercal type, made

53

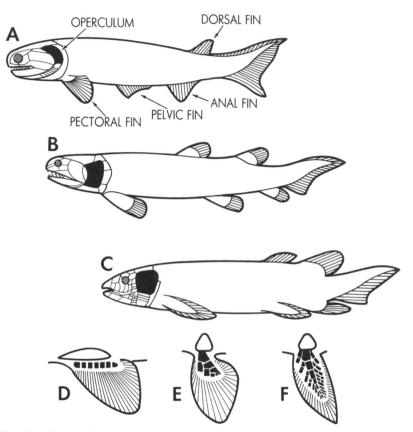

Fig. 6.1   Fin morphology and body shape of early bony and cartilaginous fishes, all about 1 foot long. *A, D, Cheirolepis,* actinopterygian osteichthyan (after Moy-Thomas and Miles, 1971; Pearson, 1982). *B, E, Osteolepis,* crossopterygian osteichthyan (after Colbert, 1980; Pearson, 1982). *C, F, Dipterus,* dipnoan osteichthyan (after Colbert, 1980; Pearson, 1982).

externally symmetrical by the development of a large ventral lobe that was stiffened internally by long fin rays.

The streamlined body form and lack of large, heavy, dermal bony plates suggests that the early osteichthyans were active swimmers, and the development of neural and haemal spines in the body musculature suggests stronger attachments for the myomeres so that the body could produce more powerful lateral undulations. The paired and median fins were well situated to act as stabilizers in the water, and in some groups the large pectoral fins had relatively narrow bases, suggesting flexibility to add fine control to turning movements. The presence of a movable operculum over the gill slits suggests the development of greater respiratory efficiency. Outward rotation of the operculum would create negative pressure in the chamber external to the gills and thus enhance water flow across the gills. The functional significance of having an ossified internal skeleton versus a cartilaginous one is not clear, and we will return to this point later.

At the time of their first appearance in the fossil record, the ostiechthyans had already differentiated into three main types, with substantial differences in fin structure, tooth type, and arrangement of dermal plates in the head region. One group, the actinopterygians (ray finned), had only one dorsal fin, and all its fins, including the paired ones, were supported internally only by fin rays (fig. 6.1). Early actinopterygians had relatively large, anteriorly placed eyes, leaving little space for the nasal capsules, which accordingly were relatively small. The jaws were relatively long, extending to the front of the head and were lined with small conical or needle-like teeth. Like the contemporary acanthodians, the early actinopterygians were probably visually oriented, active predators, feeding on a variety of smaller fish and soft-bodied invertebrates. We infer that these actinopterygians ate fish because most modern fish-eating animals have a dentition similar to these fossils. They rapidly diversified and over the past 400 million years have had an interesting evolutionary history, which we will discuss in chapter 8.

A second group of bony fishes was the dipnoans (double-breathers), also known as lungfish because their modern descendants commonly use lungs as accessory respiratory organs to their gills. The dipnoans' upper jaw was fused to the braincase, and, instead of small sharp teeth, they had a few large tooth plates with various types of low ridges and crests. There were two dorsal fins, and the paired fins were relatively long and slender and supported by an internal skeleton and musculature in addition to fin rays (fig. 6.1). The fusion of upper jaws to braincase allowed for a very powerful bite, and the shape of the tooth plates suggests optimization of crushing and chopping functions. This type of jaw apparatus suggests that the early lungfish fed on hard food items. Modern lungfish with similar tooth plates and jaws eat shellfish, snails, and crabs.

The third group of bony fishes, the crossopterygians (fringe finned), had powerful jaws lined with an outer row of small teeth, like those of the actinopterygians; in addition, they had a row of much larger, sturdy teeth on the palate. They had a movable joint between the front and back portions of the braincase and overlying dermal plates that allowed the front portion of the skull to be rotated up on the posterior part. (The other bony fishes have that braincase division early in ontogeny, but the joint becomes fused and immovable in the adults.) Like the dipnoans, the crossopterygians had two dorsal fins, and their paired fins were supported by an internal skeleton and musculature (fig. 6.1). The large teeth on the palates of the crossopterygians suggest that they fed on large fish and invertebrates, and the joint through the braincase would have allowed the front part of the skull to rotate upward and thus increase mouth gape. It was from this group of bony fishes that the land vertebrates evolved, and we will return to that story in chapter 9.

Dipnoans and crossopterygians differ from the ray-finned (actinopterygian) bony fishes in having two dorsal fins and a sturdy internal skeleton and musculature in the paired fins. In addition, they have muscular lobes at the bases of the anal and second dorsal fins and a different microstructure in the dermal plates and scales. For these reasons, the dipnoans and crossopterygians are usually classified together in a group called the sarcopterygians (fleshy finned; see fig. 4.4). Fossil members of these groups are thought to have lived in shallow bodies of fresh water or shallow, near-shore seas and to have used their muscular, paired fins not only for stabilization and control in swimming but also to help propel themselves along by pushing against the bottom.

Chondrichthyans

The last group of jawed fishes to appear in the fossil record are the chondrichthyans (meaning cartilaginous fishes). They are first known from marine deposits about 380 million years old (on about December 2). The earliest known chondrichthyans were about 3–4 feet long and had streamlined bodies covered with only tiny vestiges (called denticles) of the original bony scales. Their internal skeleton, consisting of braincase and jaws, gill arches, vertebral column (originally only neural arches), and fin supports, was composed of cartilage rather than bone, hence the name for this group. Their jaws were relatively long and lined with elongate, three-cusped teeth. They had six gill slits, each covered by a narrow flap of skin, and a small spiracle (vestigial gill slit) between the jaws and the hyoid arch. The fossil record provides no evidence that cartilagenous fishes had lungs like the placoderms and osteichthyans. They did have a regular pattern of fins: paired pectoral and pelvic fins ventrally and two median dorsal fins above, like the bony fish pattern except that the earliest chondrichthyans lacked an anal fin. The fins were supported by an internal skeleton of cartilaginous fin rays, and the paired fins were connected to the body musculature by cartilaginous pectoral and pelvic girdles. The pectoral fins appear to have had relatively broad bases where they attached to the body and to have been relatively stiff due to support from the sturdy fin rays. This is in contrast to the narrower-based, more flexible pectoral fins of some of the early bony fishes. The heterocercal tail had a large stiff ventral lobe that resulted in a symmetrical silhouette.

Their streamlined body form indicates that the earliest cartilaginous fishes were active swimmers, and they perhaps were made more efficient by the great reduction of their bony scales. The large, stiff, symmetrical tail, by analogy with that of some modern fishes, would have provided powerful thrust. The paired fins and dorsal fins were well positioned to act as stabilizers in the water, another indication of active swimming, although the broad base and stiffness of the pectorals would not have allowed these fins to be rotated for fine control of turning or braking. The long jaws and piercing teeth indicate that the early chondrichthyans were active predators, able to feed on relatively large prey.

The fast-swimming, predacious chondrichthyans known as elasmobranches (plated gills) are the ancestors of the modern sharks. Shortly after they first appeared, we have fossil evidence of another kind of cartilaginous fish, one with a shortened body flattened on the bottom and jaws lined with rows of flat, crushing teeth. These chondrichthyans are called bradyodonts (slow teeth), because their teeth were replaced more slowly than those of the elasmobranchs. From their flattened body forms and batteries of crushing teeth, we can infer that they were bottom-dwelling shellfish eaters, similar to some of the placoderms.

We have now looked at the origins of all of the major groups of fishes. For a while, they all were contemporaries. If we could go back in time and prowl through the shallow oceans, rivers, and lakes of 350–375 million years ago, we would see a marvelous diversity of vertebrates. The clumsy, heavily armored osteostracan, galeaspid and heterostracan ostracoderms, and antiarch placoderms wriggle along the bottoms, sucking up or nibbling on rich organic detritus. A great variety of flattened armored and unarmored placoderms and bradyodont chondrichthyans slowly move over the bottoms, feeding on shelled invertebrates that they crush between their broad flat tooth plates or dental batteries. In the shallow fresh waters, we would also see early lungfish (dipnoan osteichthyans) feeding on shelled invertebrates. Above the bottoms, swimming actively in the water, there are small anaspid ostracoderms and acanthodians darting at very small fish and invertebrates. The larger osteichthyans (ray-finned actinopterygians and fleshy-finned crossopterygians) and armored arthrodire placoderms are fierce predators of larger fish and invertebrates in the fresh waters, and they are joined in the seas by the elasmobranch chondrichthyans.

Of course, not all of these ancient fishes are seen together in the same places. In addition to the basic dichotomy between marine and freshwater environments, the great diversity of body forms suggests specializations for different types of aquatic habitats. Nevertheless, in many places we could see quite a great variety of fishes. In a few million years, all that will change. By 350 million years ago (December 4), all of the ostracoderms and placoderms were extinct, and the acanthodians were greatly diminished. In their places was a great diversity of bony and cartilaginous fishes, to whose stories we now turn.

Later Cartilaginous Fishes

The fossil record of the chondrichthyans, the cartilaginous fishes, consists mainly of teeth and fin spines, the only ossified parts normally found in these fishes' bodies. However, in rare cases, the cartilaginous skeleton has been preserved—sometimes even the outlines of the body and fins. Such unusual fossils have allowed us to reconstruct the series of evolutionary changes that transformed the locomotor and feeding systems of the elasmobranch chondrichthyans (the fast-swimming predators) from the primitive conditions described in chapter 6 to what we see in modern sharks. The other major group of chondrichthyans, the shellfish-eating bradyodonts, remains poorly known.

Beginning about 25–50 million years after the elasmobranchs first appeared (between December 4 and 6), paired fins with narrower bases and shorter fin rays evolved in some lines. There was also a reduction in the size and stiffness of the ventral lobe of their heterocercal tails (fig. 7.1). Narrower bases and shorter fin rays made for more flexible fins, which allowed for better control of turning movements. The smaller and more flexible ventral tail lobe provided better control over the direction of propulsive thrust from the tail. The feeding system was modified by the development of blunt, crushing teeth at the back of the jaws, while the primitive type of sharp piercing and tearing teeth was retained at the front of the mouth. This produced a multipurpose dental battery, which permitted these fishes a more varied diet (shellfish in addition to soft-bodied prey). Sharks with these modifications in fin structure and dentition, called hybodont sharks, coexisted with the primitive-type sharks, called cladodonts, for about 100 million years, until the cladodonts became extinct. Then, for another 100 million years, hybodont sharks were the dominant sharks in the oceans. Finally, about 150 million years ago (around December 20), there was another series of changes that resulted in further transformation of the locomotor and feeding systems and produced the modern sharks of today.

59

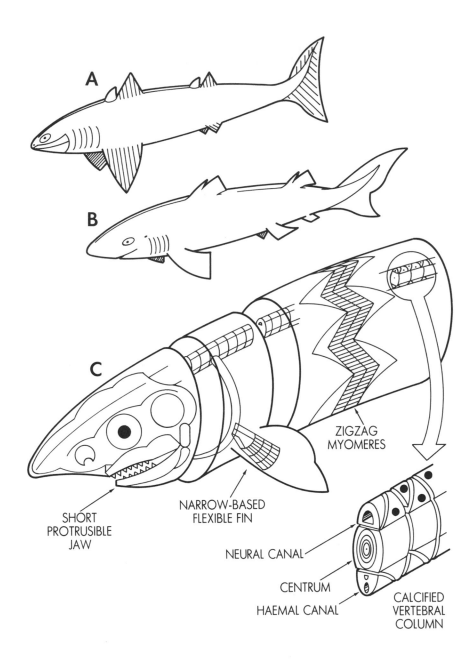

A

B

C

ZIGZAG
MYOMERES

SHORT
PROTRUSIBLE
JAW

NARROW-BASED
FLEXIBLE FIN

NEURAL CANAL

CENTRUM

HAEMAL CANAL

CALCIFIED
VERTEBRAL
COLUMN

Fig. 7.1   Evolutionary trends in sharks. *A*, *Cladoselache*, the earliest shark, 6 feet (after Moy-Thomas and Miles, 1971). *B*, modern shark body form. *C*, enlargements of head, pectoral fin, myomeres, and vertebrae, showing major changes.

The second series of changes in the locomotor system involved strengthening the vertebral column and developing a more powerful body musculature. The notochord was replaced by a chain of spool-shaped vertebral centra, to which the neural arches attached above and the haemal arches attached below. Triangular plates of cartilage filled in the spaces between adjacent neural arches and adjacent haemal arches to form a complete covering over the spinal cord above and major blood vessels below (fig. 7.1). Also, the cartilage of the vertebral column came to be impregnated with calcium salts, producing what is called calcified cartilage, a tissue similar in strength to bone. The result of these changes was a stronger axial skeleton, suggesting a more powerful body musculature and stronger, faster swimming abilities. The replacement of the continuous stiff notochord by a segmental chain of vertebral centra is thought to allow for finer control over the frequency and amplitude of lateral undulations of the body.

The way in which the body musculature was strengthened can be seen in dissections of modern sharks. The myomeres have been transformed from largely vertical, transversely oriented muscle blocks (like those seen in modern lampreys) to complexly folded, nested structures, with a zigzag profile in side view (fig. 7.1). The pull of each myomere extends over several vertebrae, allowing for more powerful lateral undulations, and the nested arrangement of the vertebrae means that many myomeres are involved in bending a given segment of the body, allowing for finer control of the movement. Finally, an anal fin appeared, suggesting enhanced stabilization capabilities.

Modification of the feeding system occurred with the appearance of shorter jaws and the development of a mechanism for jaw protrusion. Optimal design analysis tells us that shorter jaws allow for a more powerful bite. We can look at the jaws and associated musculature as a lever system. In any lever system there is a fulcrum or pivot point, two forces, and some distance from the fulcrum to the force. For a good example of a lever system, think of the seesaw at the local playground. You knew intuitively or quickly figured out that, even if you were not as heavy as your friend on the other end of the seesaw, you could balance the seesaw if you sat farther away from the fulcrum and your heavier friend sat closer. Translating that action into biomechanical terminology, you applied an in-force (your body weight) through an in-lever (the distance of your body from the fulcrum). This action generated an out-force (balanced by your friend's weight) acting some distance from the fulcrum (the out-lever length). The fact that the system was in balance or equilibrium meant that the in-force times the in-lever was equal to the out-force times the out-lever. If we solve this equation for out-force [out-force = (in-force × in-lever)/out-lever], we see that we can increase or optimize the out-force by increasing the in-force, increasing the in-lever length, or shortening the out-lever length. When we analyze muscle-bone couplings as lever systems, we can compare the relative mechanical efficiency of different morphologies. The mechanical efficiency of a lever system is a ratio of the out-force divided by the in-force. One morphology is more efficient mechanically than another if its out-force to in-force ratio is higher. The lower jaws of a shark are out-levers. The in-force of the system is the muscle that closes the jaw. By decreasing jaw length, the out-lever length is decreased and the bite or out-force is increased.

With shorter and more powerful jaws, these fishes also developed a shorter, more curved tooth row. Many modern sharks use this tooth structure to good effect with a sideways head shake that creates a sawing mechanism to cut out chunks of flesh from large prey. However, shorter jaws meant that the mouth opened behind and under the tip of the snout, an apparent disadvantage for a predator. Further, in many sharks the length of the snout was actually increased until it extended far anterior of the mouth. Perhaps in compensation for the underslung position of the short jaws, a mechanism developed to protrude the jaws forward and downward. The upper jaws lost their anterior contact with the braincase and became attached to the skull in front only by flexible ligaments. Posteriorly the jaws were linked to the braincase by the hyomandibula (the enlarged upper segment of the hyoid arch). This linkage system allowed the jaws to be protruded by swinging the lower end of the hyomandibula anterior. The retention of a long snout when the jaws were shortened, which required the development of the jaw protrusion mechanism, may reflect the need for space to accommodate large nasal capsules or perhaps to allow streamlining of the front of the head to reduce drag in fast swimming. Or, in those sharks that develop a rostrum far anterior to the nasal capsules, the large snout may have evolved to provide a surface area to accommodate the extraordinarily sensitive electroreceptor system (called the ampullae of Lorenzini). Recent work has demonstrated that sharks use this system, plus vision, olfaction, and lateral line receptors to detect and locate prey.

Coinciding with the development of protrusible jaws was the appearance of a greater diversity of tooth types, suggesting specializations for piercing, tearing, and/or slicing more efficiently than with the primitive three-cusped tooth type. Presumably the new jaw arrangement permitted elasmobranchs a greater variety of ways of attacking, killing, and eating prey, and the jaw appears to have been an important factor underlying the evolutionary radiation (begun around December 20) that produced the diversity of sharks we see today.

There are about 250–300 species of sharks alive today, including some survivors of the hybodont radiation. Most modern sharks are streamlined, fast-swimming predators with a relatively stereotyped body form (fig. 7.2). Most are between 6 and 15 feet in length, although there are small sharks, only a foot long, and giant ones (the Great White

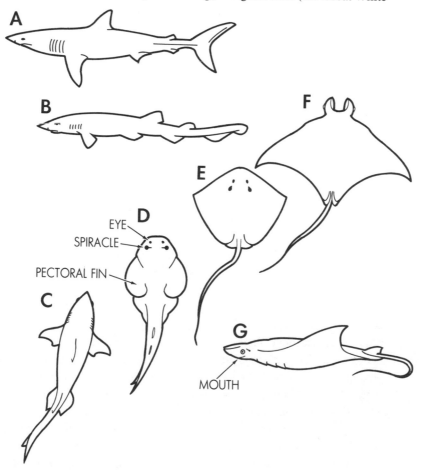

Fig. 7.2   Diversity of body form in cartilaginous fishes. *A* and *B*, sharks (after Thomson and Simanek, 1977). *C,–G,* skats and rays (after Gregory, 1951).

sharks) that reach lengths as great as 30 feet. The very largest of the sharks is the enormous (60-foot) whale shark. This animal has a reduced dentition and is a filter feeder, straining from the water the plankton, very small plants, and animals that are at the base of the ocean's food chain.

With the diversification of modern predatory sharks around 150 million years ago, one or two groups evolved specializations for living on the bottom and eating shellfish and burrowing invertebrates. These specializations involved the same sort of transformations that we saw in the bradyodonts of the first chondrichthyan radiation (most of which disappeared about 250 million years ago) and in some of the older placoderms. The body is flattened, the eyes and nostrils are shifted dorsally, the mouth is shifted ventrally, the pectoral fins are enlarged, and a battery of flat, crushing teeth is developed. In addition, the spiracle in these modern bottom dwellers is shifted dorsally and replaces the mouth as the main water intake when the fish are resting on the bottom. These kinds of chondrichthyans are called skates and rays, and we estimate that there are 350–400 species of them alive today. Most are only a few feet in length, but the giant manta ray reaches widths of up to 15 feet. Like the giant whale shark, it too is a plankton filter feeder. Fortunately, among the skates and rays alive today we have animals that preserve various stages in the modification of body form from typical streamlined shark to most flattened ray (fig. 7.2). This kind of morphological series allows us to visualize how the evolutionary transformations could have occurred. The most flattened types can no longer swim by lateral undulations of their body. Instead, they develop thrust by vertical undulations of their greatly expanded pectoral fins, or in some cases by flapping the pectoral fins like the wings of birds.

Until recently, sharks were considered more primitive than bony fishes, largely because they have a cartilaginous skeleton. Cartilage appears before bone in the ontogenetic development of most vertebrates. Simplistically applying the notion that ontogeny recapitulates phylogeny, many biologists concluded that therefore the cartilaginous fishes must be more primitive than bony fishes. Recent phylogenetic analyses such as that shown in this book demonstrate, however, that the cartilaginous adult skeleton is a secondarily derived feature in Chondrichthyes. Bone is known in other groups of early fishes such as the Ostracoderms and Placoderms. The correct explanation of the distribution of a cartilaginous skeleton in sharks is that bone was present in the species that gave rise to the first sharks but was lost during the further evolution of the group.

Simple notions of *scala naturae* considered cartilage to be a tissue inferior to bone. We now know from mechanical analysis that cartilage is not inferior, but different. Cartilage is a better tissue than bone for rapid growth and therefore a better material for laying out the framework of the internal skeleton in fast-growing embryos. Also, the calcified cartilage that makes up the internal skelton of adult sharks seems to be an alternative, but equally efficient, type of supporting structure, biomechanically comparable with the ossified skeleton of the bony fishes. The simplified notions about the primitiveness of sharks also led biologists to believe that sharks were inferior to bony fishes in their behavior and brain anatomy as well. Recent studies have shown, however, that shark behavior is considerably more complex than we previously imagined and, most surprising, that some kinds of sharks have relatively larger brains than any bony fish. Some, in fact, attain a relative brain size as large as that found in birds and mammals. Thus there is no reason to consider sharks as more primitive than or "inferior" to the bony fishes.

Until now we have talked about the descendants of the elasmobranch branch of the early chondrichthyans. What about the bradyodont radiation in the early days of the cartilaginous fishes? There is a small group of chondrichthyans known as chimaeras or ratfishes (or scientifically as holocephalans), which may be the living descendants of bradyodonts. They have crushing toothplates, upper jaws fused to their braincase, an operculum over the gill slits, enlarged pectoral fins, and a skinny whip-like tail. Because most chimaeras are deep-water marine fishes, it has been difficult to study them and we still know little of their habits and life histories. There are about twenty-five known species of living chimaeras.

Later Bony Fishes

The history of the bony fishes (osteichthyans) is predominantly a story of the evolutionary transformations and radiations of the actinopterygians, the ray-finned bony fishes. For about 150 million years, there was little change in the basic body organization described in chapter 6. There was diversification based on size, with species ranging from a few inches to several feet in length, and on differences in body shape, with outlines ranging from long and thin to short and deep bodied. However, it was not until about 250 million years ago (around December 12) that the ray-finned fishes began a series of modifications of their locomotor and feeding systems that resulted in the extraordinary diversity of bony fishes that we see around us today.

The changes in locomotor system included strengthening the axial skeleton and body musculature, modifying the tail shape, developing more flexible and more versatile fins, thinning the body scales, conversion of the lungs to a swim bladder, and shifting position of the paired fins (fig. 8.1). Ossified vertebrae replaced the elastic notochord, and long neural and haemal spines developed in the median partition between the left and right rows of myomeres. Further, ribs developed between successive myomeres in two places: at the intersection of the transverse partitions with the horizontal partition and within the connective tissue sheet around the viscera (abdominal cavity). These changes, which resulted in a more strongly braced skeleton for muscle attachment and for transmission of thrust from the tail, are correlated with the development of a stronger body musculature. The myomeres of modern bony fish, like those of modern sharks, are complexly folded into a zigzag, nested pattern, so that the pull of a given myomere is exerted over several vertebral segments and several myomeres are involved in bending a given part of the body. This arrangement of the myomeres relative to the vertebrae allows stronger and more controlled lateral undulations. The replacement of notochord by a chain of short vertebrae

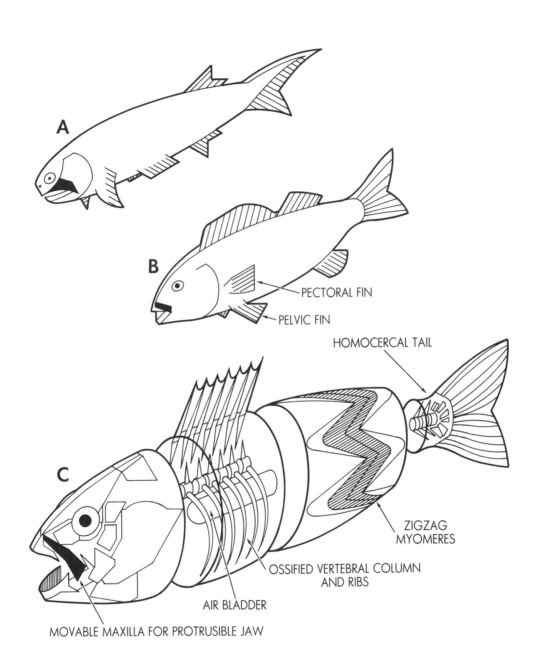

A

B

PECTORAL FIN

PELVIC FIN

HOMOCERCAL TAIL

C

ZIGZAG
MYOMERES

OSSIFIED VERTEBRAL COLUMN
AND RIBS

AIR BLADDER

MOVABLE MAXILLA FOR PROTRUSIBLE JAW

Fig. 8.1 Evolutionary trends in teleost bony fish. *A*, *Cheirolepis*, an early actinopterygian bony fish, 1 foot. *B*, perch, a modern teleost. *C*, enlargements of head and body
showing major changes.

also allows for finer control over the frequency and amplitude of lateral undulations, particularly in the tail region.

The primitive heterocercal tail was modified by shortening of the vertebral column so that it no longer extended into the dorsal part of the tail fin. Instead, it terminated in a few broad plates at the base of the tail, and from these plates flexible fin rays radiated out to support a deep symmetrical tail (fig. 8.1). This is called a homocercal tail and is thought to allow finer control over the intensity and direction of propulsive thrust by subtle modifications of tail fin shape. The fin rays were reduced in number and stiffness in the other fins as well, and a musculoskeletal system that provided greater control over fin shape during swimming movements also developed.

The body scales lost their original outer layers of enamel and dentine-like tissues and their bony bases thinned out, resulting in the very light and flexible scales seen in most ray-finned fishes today. This change improved locomotor abilities by making the body lighter and more flexible. The conversion of lungs to swim bladder allowed fishes better regulation of buoyancy, by the addition or removal of gas from the bladder through a network of blood vessels. Thus the fishes could hover at different depths with less expenditure of energy through muscular activity. The development of a swim bladder was accompanied by a change in the location of the paired fins. The pectoral fins shifted dorsally, to a position at about midlevel on the side of the fish, just posterior to the operculum, and the pelvic fins moved anterior to the original position of the pectorals, ventral to the operculum.

Behavioral studies show that a major function of the raised pectoral fins is braking. Mathematical analysis indicates that, as the pectoral fins brake, they produce lift; yet fish do not rise in the water column when using their pectoral fins to brake. Therefore something must be countering this force. When the anteriorly placed pelvic fins are removed, fish do rise in the water while braking. It therefore appears that, as the pectoral fins moved dorsally and evolved a braking function, it became necessary to move the pelvic fins forward on the body to minimize lift or pitch.

So far we have talked about drag being the major problem in body design with respect to fish locomotion. We have shown that the optimal shape for minimizing drag during sustained locomotion that is powered by the caudal fin is a streamlined, dorso-laterally flattened body shape. But if this is the case, why are all fish not streamlined, and why did many teleosts evolve median and paired fin locomotion? The answer is that not all fish swim at high speeds or live in open water. Many fish live in and around coral reefs and other structurally complex habitats, swimming relatively short distances. And the optimal design for slow swimming and precise maneuvering is not the same as the optimal design for high-speed, sustained swimming. Experimental studies show that, at low swimming speeds, median and paired fin propulsion are energetically more efficient than body or caudal fin propulsion. Also, median and paired fins can usually function independently of each other and thus can precisely orient thrust in any direction, producing a much finer control of movements than is possible using only caudal fin propulsion. The evolution of such features as midlateral pectoral fin insertion; ventrolateral pelvic fin insertion; soft-rayed, dorso-ventrally symmetrical dorsal and anal fins; and short, deep bodies made possible a greater diversity of swimming modes, including hovering and precise maneuvering, than was ever possible with a propulsive system dominated by the tail fin. Diversification of fin shape and position is in part responsible for the extraordinary variety of body shapes seen among modern actinopterygians (fig. 8.2).

The changes in the feeding system included shortening of the jaws and uncoupling the upper jaw from the operculum bones behind it. Bones, ligaments, and muscles were modified to allow for rapid expansion of the mouth cavity and, in advanced forms, protrusion of the jaws (fig. 8.3).

This has been a difficult system to study because of the small size of the anatomical components and the extremely high speed at which they operate. It takes some species only 5–10 milliseconds to go from closed jaws to full gape. However, in the past fifteen years, refined techniques of functional anatomy have provided insights into how the system works. Electromyography allowed us to see the sequence of muscle action in relevant head muscles. Cineradiography (x-ray movies) at very high speeds showed how the bones moved. Pressure gauges recorded changes in pressure in the mouth and pharyngeal chambers as the fishes fed and respired. We now know that mouth opening involves three musculoskeletal couplings (fig. 8.3). The head is elevated by a muscle attached to the back of the skull. There are two pathways for opening the lower jaw. In the first, a muscle attached between the operculum and the skull pivots the lower jaw ventrally and posteriorly through the action of a ligamentous attachment between the operculum and lower jaw (labeled 2 in fig. 8.3). In the second path, the lower jaw is opened by a

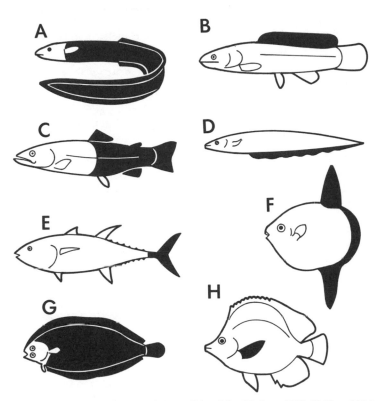

Fig. 8.2 Diversity of body shapes in bony fishes (after Lindsey, 1978; Webb and Blake, 1985; Young, 1962). Black areas indicate which parts of the body and fins are involved in propulsion. *A, Angilla* (eel). *B, Amia* (bowfin). *C, Salmo* (salmon). *D, Gymnotus* (knife fish). *E, Thunnus* (tuna). *F, Mola* (sunfish). *G, Solea* (flounder). *H, Chaetodon* (butterfly fish).

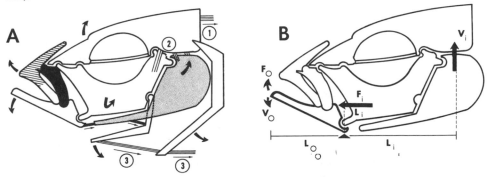

Fig. 8.3 Diagrammatic bony fish skull (after Liem, 1978). *A*, numbers indicate the three pathways of mouth opening. Arrows show direction of bone movement. The maxilla is colored black. The premaxilla is striped, and the operculum is lightly shaded. *B*, two important lever systems in jaw opening and closing. F_i shows the position of the major jaw closing muscle. V_i shows the position of a major jaw opening muscle.

posterior pull of the operculum. This movement is controlled by muscles attached to the hyoid and pectoral girdle (labeled 3 in fig. 8.3).

When the mouth opens and the volume of the mouth chamber is
rapidly expanded, strong suction is created in the area just in front of the
mouth, providing a new (among osteichthyans) type of food ingestion:
suction feeding. This transformation of the feeding apparatus resulted in
what is undoubtedly the most complex bone-joint-muscle system in all
the vertebrates and provided the anatomical basis for an extraordinary
diversity of mouth parts and feeding specializations among the modern
ray-finned fishes.

Biomechanical analyses based on lever system relationships show us
that the diversity of fish skull morphology (fig. 8.4) is closely correlated
with the food preferences of the various species. Fish specializing in
scraping food off rocks have maximized the force of jaw closing. Fish
specializing in eating fast-moving prey (usually other fish) have maximized the speed of jaw opening. Consider the jaw-opening and -closing
systems of a fish as a series of lever arms (fig. 8.3). Remember from
chapter 7 that the out-force (in this case the force of jaw closing) is
maximized by having a long in-lever and/or a short out-lever length.
Look at the effective length of the lower jaw (the horizontal straight-line
distance from the fulcrum to the tip of the jaw) in B and C in figure 8.4
relative to fish A, a more generalized feeder. In both food scrapers the
out-lever has been shortened. Conversely, the speed of jaw opening
(V_o) is maximized when the out-lever is lengthened or the in-lever
shortened. (This is because, at equilibrium, $V_oL_i = V_iL_o$.) The elongate
jaws of D and E (relative to the jaws of A) show that the out-lever arm
has been lengthened in fish that feed on fast-moving prey.

The morphological changes associated with improved suction feeding also enhanced respiration. With closure of the mouth, constriction
of the mouth chamber and opening of the operculum, water is drawn
across the gills in a respiratory current that can be controlled by modification of gill shape.

These evolutionary changes in the locomotor and feeding system did
not happen all at once, and they apparently occurred in different lines of
ray-finned fishes at different times. Actinopterygians may be divided
into three grades of structural organization, based on the degree to
which they show the changes outlined above: chondrostean, holostean,
and teleostean. The most primitive level, the chondrostean grade, includes the earliest ray-fins and was the dominant form for the first 200

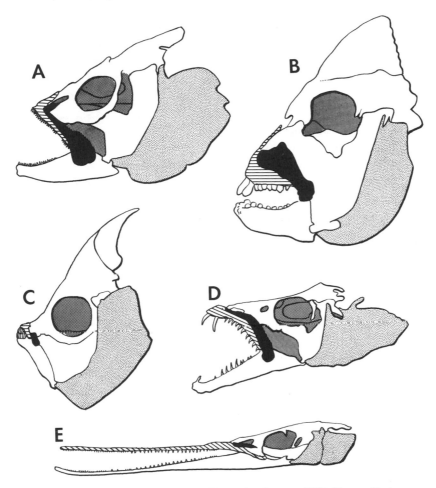

Fig. 8.4   Diversity of skull shape in teleost fishes (after Gregory, 1959). The maxilla is colored black. The premaxilla is striped, and the operculum is lightly shaded. *A*, small mouth bass (*Micropterus dolomieu*). *B*, sheep's head (*Archosargus probatocephalus*). *C*, spot fin butterfly fish (*Chaetodon ocellatus*). *D*, cutlass fish (*Trichiurus lepturus*). *E*, needlefish (*Tylosurus marinus*).

million years of actinopterygian history. There are a few chondrostean survivors alive today, like the sturgeon and paddlefish, freshwater forms that are somewhat modified from the primitive chondrostean condition. The holostean grade predominated from about 200 to 130 million years ago (or from December 16 to 21 or our time-scale year), and during that time they underwent major evolutionary radiations that produced much of the same diversity of body forms as occurred in the earlier chondrostean radiations. Today the only holostean survivors are the bowfin and gar. Teleost fishes first appeared around 135 million years ago (around December 21). They soon replaced the holostean types and began an explosive evolutionary radiation that produced the greatest diversity of species of any group of vertebrates that ever evolved.

We estimate that there are about 23,000 species of teleosts alive today, a number that accounts for more than half of all the living species of vertebrates. They range from tiny fishes less than an inch long to giants longer than 15 feet, and they inhabit virtually all the waters of the earth, from the deep ocean depths to high mountain streams. They utilize virtually all imaginable food resources, and among them are predators on a great diversity of vertebrates and invertebrates, filter feeders, herbivores, omnivores, and parasites. In keeping with their diversity of diets and habitats, teleosts display an extraordinary variety of behaviors and life histories and an almost unbelievable diversity of body forms.

In contrast to actinopterygians, whose evolutionary history culminated in the great teleost radiations, the other two groups of bony fishes played only modest roles in the world's aquatic communities. Shortly after the lungfishes (dipnoan sarcopterygians) first appeared, their vertebral column changed orientation so that it projected horizontally into a symmetrical tail. Additionally, the dorsal and ventral median fins were displaced posteriorly so that they formed one long, continuous, flexible fin around the tail. Lungfish retained this basic body form throughout their evolutionary history and never evolved much diversity in their feeding or locomotor systems. There are three genera surviving today, one each in South America, Africa, and Australia. They live in shallow, fresh-water lakes and rivers, move slowly along the bottom, and use their powerful jaws and tooth plates to chop and crush shellfish and plant material. All the living lungfish can use their lungs to breathe air but vary considerably in the degree to which they do so. The African genus takes about 90% of its oxygen from the air, while the Australian lungfish takes almost all of its oxygen from the water. If their lake or pond dries up, the African and South American genera can burrow into the mud and survive in a state of almost suspended animation (aestivation) for as long as several years. The Australian form lives where the water never dries up. Fossil lungfish burrows are known from deposits around 300 million years old, so this ability to survive droughts is quite an ancient adaptation.

The other group of sarcopterygians, the crossopterygians, underwent a small evolutionary radiation as medium-sized, fresh-water predators that soon became extinct except for two lines. One group, called coelacanths, moved into the oceans and played a minor role in the marine bony fish communities. They can be traced in the fossil record until about 65 million years ago, when, it is thought, they became extinct. Then, in 1938, a living coelacanth was caught in deep waters off the island of Madagascar, providing us with the first opportunity to study the soft anatomy of a crossopterygian fish. Many more coelacanths have been brought up in the past 45 years, but, because they are deep-water fish and have not been maintained in captivity, not much is known about their lives and behavior. From the rhipidistians, the other group of crossopterygians (fig. 4.4), one line survived extinction and underwent the most extraordinary evolutionary transformation that one might imagine for a fish: it evolved into the first land vertebrates. The story of these descendants will occupy the remainder of this book.

Designs for Land Life

Between 370 and 360 million years ago (December 2–3) in the midst of the great evolutionary radiations of jawed fishes, one species of fleshy-finned (sarcopterygian) bony fishes began one of the most important evolutionary transformations in the history of vertebrates: it evolved the ability to live on land. The transition from an aquatic life to a terrestrial one required modifications of several important body systems to solve problems of support and locomotion, respiration, dessication (drying out), reproduction, feeding, and the functioning of some sensory systems. The reorganization of these body systems is related primarily to the different physical properties of air and water. The density of water is 1 g/cm^3 while that of air is .001 g/cm^3. For purposes of comparison, muscle has a density of 1.05 g/cm^3 and that of bone is about 3 g/cm^3. In this chapter, we will discuss the features that evolved in a crossopterygian fish to meet the requirements for life on land in relation to this difference between water and air.

Support against the pull of gravity is a minimal problem for an aquatic animal because the body is not much denser than water and because it is buoyed up by the water. However, an animal the size of a crossopterygian fish (2 or 3 feet long) requires a substantial support system to elevate the body off the ground in air. Such elevation is necessary to keep the lungs from being crushed by the weight of the body and, as we will see below, for locomotion.

Body support was achieved on land by modification of the paired fins, pelvic and pectoral girdles, and vertebral column (fig. 9.1). In crossopterygians, the pectoral and pelvic fins had a sturdy internal skeleton and musculature that inserted close to the body. These fins were enlarged and transformed in the first land animals into short, powerful front and hind limbs (fig. 9.2). Both pairs of limbs had a similar design, and land vertebrates are generally referred to as tetrapods (four footed). Proximally there was a single bone (the humerus in the forelimb, the

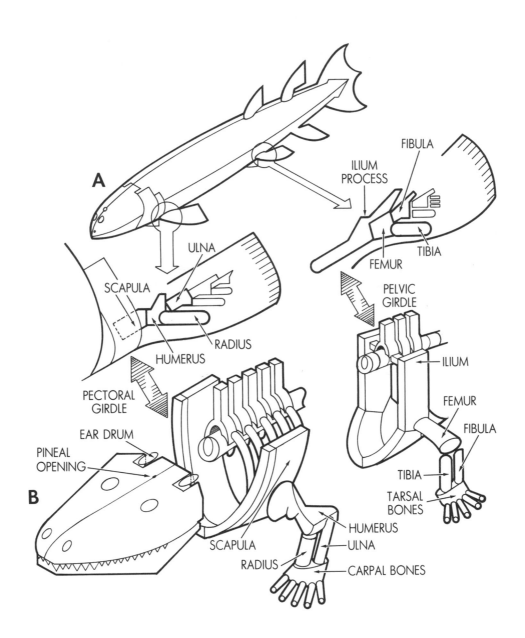

Fig. 9.1   Designs for land life. *A*, *Eusthenopteron*, a crossopterygian fish, with enlargements of pectoral and pelvic fin structure. *B*, head, forelimb, and hindlimb of early labrinthodont amphibian.

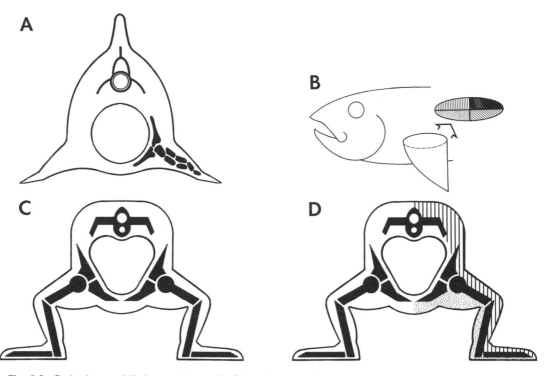

Fig. 9.2 Body shape and limb musculature of a fish and a tetrapod. *A*, cross section through the body of a fish at the level of the pectoral fins. *B*, the dorsal and ventral muscles of the paired fin of a fish. *C*, cross section through the body of an early tetrapod at the level of the forelimb. *D*, the flexor and extensor musculature of the limb of a tetrapod.

femur in the hindlimb); then more distally two bones (radius/ulna and tibia/fibula); then a zone of several small bones (carpals and tarsals) forming wrist and ankle regions; then a fan shaped array of five stubby bones (metacarpals and metatarsals), each of which terminated in a line of two to five bones (phalanges) that constitute the digits (fingers and toes). The first two sets of bones (humerus/femur and radius-ulna/tibia-fibula) can be traced to a similar pattern of bones in the paired fins of the ancestral crossopterygians, but the more distal elements appear to be a new design, replacing the peripheral bones and fin rays of the ancestral fins.

The pectoral and pelvic girdles were expanded and strengthened to allow the limbs to support the weight of the body. The pectoral girdle lost its connection to the back of the skull and developed a fan-shaped sling of muscles on its inner surface within which the body was sus-

pended (via the rib cage). The loss of the attachment between the pectoral girdle and the head meant that the head could now move
independently of the body. The pelvic girdle, which was relatively
small and embedded in the body wall musculature in crossopterygians,
expanded upward to make contact with the vertebral column via a pair
of short ribs. Thus a direct bony connection was made between the
hindlimb and the vertebral column. As we will see below, such a connection is important for transmitting the propulsive thrust of the
hindlimbs to the body. The lower portions of both the pectoral and
pelvic girdles were greatly enlarged to provide attachment area for
powerful muscles that extended out the undersides of the limbs; when
contracted these muscles raised the body off the ground. Other muscles
extended from the upper portions of the girdles out to the upper surfaces
of the limbs and helped pull first the limbs and then the body forward
during walking.

Finally, the vertebral column was strengthened by the development
of interlocking processes (called zygapophyses) between adjacent neural arches, and by the eventual replacement of the notochord by one or
two pairs of bony rings and/or nodules (centrum elements) seen around
the notochord in the crossopterygian fishes (fig. 9.3). The addition of
zygapophyses limited the amount of bending of the backbone in the
dorso-ventral direction. There is a substantial loading on the vertebral
column in a terrestrial animal because of the gravitational force on the

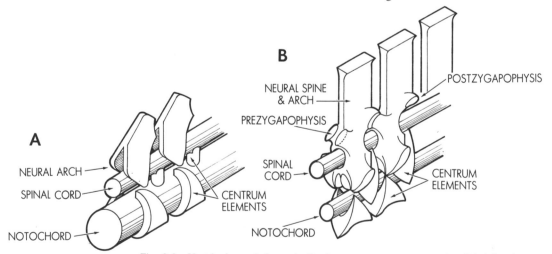

Fig. 9.3 Vertebral morphology. *A, Eusthenopteron,* a crossopterygian fish (after Andrews and Westoll, 1970). *B, Eryops,* a labrinthodont amphibian (after Gregory, 1951).

body mass, and these processes help to provide support to prevent the belly from sagging toward the ground. Additionally, the zygapophyses restricted the extent to which the backbone could twist around its long axis during terrestrial locomotion.

A series of remarkably well preserved fossils allows us to trace these evolutionary transformations. The bones are preserved, and many of them show scars and patterns where fin and limb muscles were attached. By looking at these fossil remains and the structure of modern tetrapods, or land vertebrates, we can infer the muscle transformations that were involved. By their consistent position and similar nerve innervation pattern, it is clear that the same muscles that control paired fin movement in fishes have been modified to support the body and move the tetrapod limb (fig. 9.2). The dorsal muscles of the fish's paired fin have become the extensors of the tetrapod limb, and the ventral muscles of the paired fin are the flexors of the tetrapod limb. We can reconstruct the position of the limbs relative to the body in fossil animals because the articular (end) surfaces of adjacent limb bones and joints of girdles into which they insert often fit together only when bones are oriented in a particular direction. Also, depending on the degree to which the articulation forms a ball-and-socket type of joint, directional movement of the limbs may be constrained or impossible.

In most fish, propulsive thrust is provided by the axial musculature of the body wall and the caudal fin. The contracting myomeres send the body of a fish into lateral undulations that, together with the caudal fin, push against the dense medium and cause the animal to move foward. The paired fins are used primarily for stability, braking, and maneuverability (see chap. 7). As a result of this form of locomotion, the body wall musculature of a fish is thick and large, while the musculature controlling the movement of the paired fins, attaching proximally on the fin, is relatively small (fig. 9.2). On land, the tail fin cannot be used as a source of propulsive thrust because the air is not dense enough. And, as we mentioned earlier, the pull of gravity on the body is a much greater problem in air than in water, and the lungs will be crushed unless the body is supported off the ground or there is some protective casing for the lungs. In the first land animals, the paired fins were modified to support the body. Once the body was supported by the paired appendages, these limbs were recruited into the locomotory system. The short stubby limbs were held out laterally with their proximal segments (humerus and femur) horizontal and perpendicular to the body. This type of limb position is often referred to as a sprawling posture (fig. 9.2). Lateral undulations of the body threw first one foot and then the other forward. It is likely that the footfall pattern of the earliest land animal was right front, left hind, left front, and right hind. It is this sequence of foot movements that allows a tetrapod to be stable during locomotion with only three feet on the ground at a time. We know that the body was supported during locomotion and that the feet were oriented forward because the tracks or early amphibians walking in mud have been preserved and dated.

Lateral undulations were not, however, the only source of propulsion. In addition, the newly developed musculature extending from the body axis and girdles to the limbs rotated the limbs forward and pulled the body forward on the limbs. A special set of muscles from girdles to humerus and femur rotated those short wide bones about their own horizontally oriented long axes. That motion, called long bone rotation, added to the length of the step. Finally, two sets of muscles extending from the distal ends of the humerus and femur to the hands and feet flexed and extended the wrist and ankle joints in propulsive and recovery strokes. In modern salamanders who have body proportions roughly similar to those of early tetrapods, lateral undulation accounts for about 18% of a step length or the distance a foot moves from where it is lifted off the ground to where it next touches the ground. In comparison, limb retraction accounts for over 50% and long bone rotation close to 30% of a stride length. Not unexpectedly, as the body wall musculature decreased in importance and size in tetrapods, the limb controlling muscles enlarged and evolved a more distal insertion on the limb (fig. 9.2).

The gills of fish and crossopterygians have no internal skeletal support. It is not necessary because the animals live in water and the gills float in the water. In air, the surface tension of wetted gill surfaces makes the filaments stick together, reducing total respiratory surface for gas exchange drastically. This is the reason a fish will suffocate on land, even though air has more oxygen in it per unit volume than does water. This and the problem of water evaporation in air mean that gills are not a reasonable respiratory organ for land animals. The earliest land animals lost their gills in the adult stage and relied instead on lungs for breathing. This was not a major transformation of the lungs. They were used as respiratory organs in the crossopterygians, the tetrapod's fish ancestors. With life on land, however, there was an increase in the numbers of blood vessels in the lung walls.

Land animals also helped solve the problem of desiccation, or too much water loss, by using internal structures or lungs for gas exchange. But the enclosed lungs in the body also posed a new problem—how to get the air to the internally situated lungs. A large, active land vertebrate needs some sort of pumping system to transport air into and out of the lungs. One group of early tetrapods developed a rib cage suction pump. This device used the body wall musculature in which the ribs are embedded to alternately constrict and expand the body cavity around the attached lungs. Expansion of the body cavity and lungs creates a pressure differential, air is sucked into the lungs, and contraction of the body cavity expels the air out of the lungs.

Another group of early land animals solved the problem a different way. They lost their body scales, evolved to a small size, and relied on gas exchange across their thin, moist skin to supplement lung breathing. With skin respiration the amount of oxygen exchanged is a function of the surface area of the body over which diffusion can take place. There must be skin (surface area) enough to support the oxygen demands of the body cells (volume) of the animal. Smaller objects have a higher surface area to volume ratio than do larger objects. The reason is that, as an object gets bigger, the area increases as a square and the volume increases as a cube (think of the formula for the area of a sphere. $A = 4\pi r^2$, compared to the volume of a sphere, $V = 4/3\pi r^3$). As a result of this scaling relationship between volume and area (cube versus square), larger animals do not have enough surface area for gas exchange in relation to the volume of body tissue that needs oxygen. One of the limitations of using the skin as a major site of gas exchange is absolute size. Salamanders and frogs are living animals that use both skin and lung respiration, and they are fairly small animals that are restricted to moist environments.

Desiccation, or drying out, is generally a large problem for land vertebrates once they leave very moist environments. The earliest tetrapods inherited from their fish ancestors a complete covering of dermal scales, which required little modification to protect the body from drying out. However, the eyes and olfactory receptors could not be protected from desiccation by dermal scales; to provide them with a constant covering of moisture, tear or lacrimal glands and lids evolved.

The lateral line system is the major sensory system that detects vibrations in the water surrounding the fishes. This system does not work on land because the air is not dense enough to cause the hair cells to move.

Therefore a new system evolved to allow the early land vertebrates to detect airborne vibrations.

When sound travels through some medium, it makes the medium vibrate and causes changes in pressure. The ratio of the size or amplitude of the pressure change divided by the size or amplitude of the velocity change is called the acoustic impedance of the medium. When sound travels from one medium to another, the degree to which the sound is transmitted depends on how close the acoustic impedances of the two media are. "Hearing" in fish is accomplished by deformation of special processes of hair cells located inside the otic capsule (inner ear) of the fish. There is no connecting channel from the inner ear to the outside of the body, but sound travels from water through the body of the fish to the inner ear as though the body were transparent. This is because water and the fish body have similar acoustic impedances, in part because of their similar densities. On the other hand, 99% of the sound energy traveling from air to the body of an organism is reflected rather than transmitted because of the difference in acoustic impedance between air and the body. (Remember the difference in density between air and muscles and bones.)

The transition from water to land meant that there had to be some change in the hearing apparatus if the first terrestrial animals were to pick up sound. Two changes took place. First, a fluid-filled perilymphatic system evolved in the inner ear to improve how effectively small displacements in the tissues of the head could be transmitted to the hair cells in the inner ear. We infer that such a perilymphatic system was present in the earliest tetrapods because all of their living descendants (modern amphibians, reptiles, birds, and mammals) have a fluid-filled perilymphatic system.

Second, the hyomandibula (the enlarged upper segment of the hyoid arch that had become a jaw brace in early fishes) was further modified to transmit vibrations from the outer surface of the head into the perilymphatic system of the inner ear. The hyomandibula was shifted in position to form a strut between the inner ear (located in the otic capsule at the back of the braincase) and the side of the head. In its new function as sound vibration transmitter, the hyomandibula acquires a new name, the stapes, and the region in which it is enclosed is called the middle ear. At a later time, in some land animals, a round, flexible membrane called an eardrum or tympanic membrane developed in a notch at the back of the skull. The eardrum and stapes function together to increase the sensitivity of the ear to sound traveling in air by improving the acoustic impedance match between the air and the body. Judging from middle ear anatomy in living and fossil tetrapods, the evolution of a tympanum-stapes sound transmitting system occurred at least three times independently. It is found in some early tetrapods, while in others the stapes alone was present. In some of these latter groups there is never a trace of a notch in the back of the skull for a tympanic membrane, and, like the earliest tetrapods, they picked up sound vibrations that were transmitted via the stapes to the inner ear through the soft tissues and bones of the head.

Reproduction in most fishes is highly dependent on water. Sperm and eggs are usually shed into the water and fertilization takes place outside the body of the female (external fertilization). The fertilized eggs complete their development in water. Reproduction on land presents obstacles to external fertilization and developing embryos. In both cases, desiccation will occur unless the land habitat is extremely moist. Two solutions to the problem of drying out in land reproduction are internal fertilization and the formation of some water barrier between the developing embryo and the drying air. These two solutions did not evolve in tandem until at least 50 million years after the first amphibians appeared, and their evolution marks the advent of the reptiles (see chap. 11). The earliest land vertebrates, the amphibians, probably returned to the water for reproduction. Some of their descendants, the modern frogs and salamanders, remain restricted to moist habitats and water for reproduction (see chap. 10).

Prey capture in fishes is generally accomplished by suction feeding. The fish rapidly expands its mouth cavity, sucking in water and whatever food is suspended in the water column. Suction feeding is impossible in air. Prey are much denser than the air and do not stay suspended but are rapidly pulled down by gravity. It would take an incredibly fast working, high-powered suction system to be able to pull in airborne prey by increasing the volume of the mouth cavity, and thus the transition from water to land required new mechanisms for prey capture. It is with the advent of land living that the tongue appears in vertebrates. It is derived from parts of the gill bar apparatus and head myotomal musculature.

Given all the problems that had to be overcome for successful life on land, we might wonder why a group of crossopterygian fishes got involved in that difficult transition in the first place. In other words, what were the selective pressures that made it advantageous for a large predacious fish to venture onto land? An early answer to this question was influenced by the fact that the transition took place during a time when many areas of Earth were experiencing periods of unusual aridity. This led to the suggestion that our crossopterygian ancestors first ventured out onto land to crawl from a shrinking body of water to a larger one. We can regard the ability of modern lungfish to burrow into the mud and aestivate under similar conditions as an alternative solution to the same problem.

However, it seems likely that the transition from aquatic to land life would probably have taken place in a relatively moist environment, where it would have been easier for essentially aquatic animals to survive than in a harsh, dry habitat. It has been suggested that crossopterygian fishes moved out onto the land to avoid being eaten by other fishes. Although the crossopterygians were the largest predators in their communities, in their larval or juvenile stages they would have been suitable prey for other predators or for adults of their own kind. Another suggestion is that the land offered a rich source of animal food, in the form of a great diversity of arthropods (insects and their relatives, some of which were quite large), and little competition from other predators. Whatever the reason, we know from the fossil record that the transition took place by around 360 million years ago (or around December 3 on our time-scale year). Shortly after that time, the earliest land vertebrates, the amphibians, had begun their evolutionary radiations into the great diversity of ecological niches available on land.

The Amphibians

The oldest known land vertebrates are extremely primitive amphibians whose fossil remains have been found in deposits around 360 million years old. They were about 2–3 feet long, about the size of their crossopterygian fish ancestors, and they had completed the transformation of paired fins to limbs and had lost their internal gill apparatus as adults, indicating that they breathed air and could move about on land. However, these early amphibians, known as ichthyostegids, retained a tail fin (indicated by the presence of fin rays) and canals for the lateral line system on the skull. These two features suggest that they spent most of their time in the water. Replacement of the notochord by bony vertebral centra was not yet completed, and beneath each neural arch was a half-ring of bone plus a pair of small nodules around the notochord. Perhaps because they lacked a strong vertebral column, the ichthyostegids had expanded, overlapping ribs. The enlarged ribs would have provided a strong, rigid rib cage to support the viscera. Their jaws were lined with large, sharp, conical teeth, indicating that the ichthyostegids were predators, probably on large fish. Details of their skull anatomy suggest that the ichthyostegids were an early offshoot from the main line of amphibian evolution and not ancestral to later amphibians (fig. 10.1).

Between 340 and 250 million years ago (December 5–11), there was a major evolutionary radiation of amphibians. The various lineages classically have been divided into two main groups, based primarily on differences in body size, skull bone patterns, and vertebral column anatomy: the large, powerfully built labyrinthodonts and the small, slender lepospondyls. This is now considered to be an artificial division (fig. 10.1) that does not reflect true phylogenetic relationships; nevertheless, it is convenient for introducing the different types of early amphibians.

The early labyrinthodonts were stocky, short-legged, large-headed animals, most of which were between 2 and 4 feet long. Their skulls

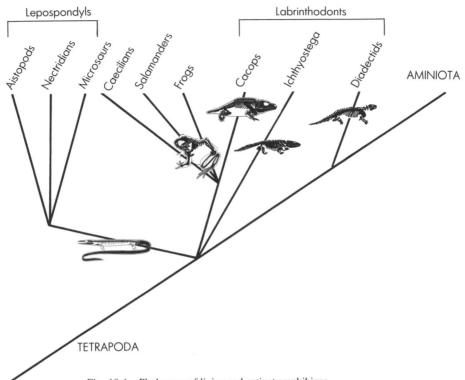

Lepospondyls

Labrinthodonts

Aistopods

Nectridians

Microsaurs

Caecilians

Salamanders

Frogs

Cacops

Ichthyostega

Diadectids

AMINIOTA

TETRAPODA

Fig. 10.1   Phylogeny of living and extinct amphibians.

were moderately deep and massive, and their long jaws were lined with small, sharp, conical teeth. In addition, many species had a second tooth row, consisting of a small number of large teeth on the roof of the mouth. The teeth were covered with a layer of infolded enamel, like those of their crossopterygian ancestors, and this characteristic gives them the name labyrinthodont (labyrinth toothed). Many labyrinthodonts had a notch at the back of the skull that accommodated a large eardrum and a sturdy stapes that transmitted vibrations from the eardrum to the otic capsule that enclosed the inner ear. The vertebrae consisted of a neural arch above and usually two pairs of centrum elements below, around the vestiges of an increasingly constricted notochord. There was considerable diversity in the relative sizes of the centrum bones and the degree to which they interlocked. These variations may reflect evolutionary experimentation in developing a stronger vertebral column, which was needed to support body weight on land, while main-

taining the lateral flexibility used in locomotion. Bony dermal scales
covered the body, and a sturdy rib cage, with its musculature, probably
functioned as a suction pump for lung respiration.

The relatively ponderous skeleton and short limbs (fig. 10.2) suggest
that the majority of the labyrinthodonts were slow, clumsy walkers on
land, and they probably were truly amphibious and spent much of their
time lying about in shallow water, like modern crocodiles. The long

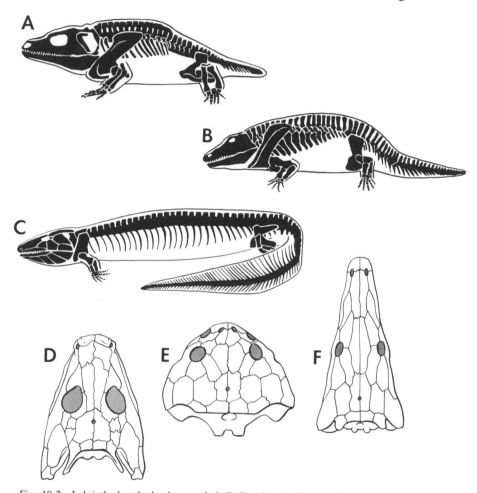

Fig. 10.2   Labrinthodont body shape and skull diversity. *A, Cacops,* about 16 inches
long (after Romer, 1966). *B, Eryops,* about 5 feet long (after Gregory, 1951). *C,
Eogyrinus,* about 10 feet long (after Romer, 1966). *D, Palaeogyrinus,* representing the
primitive skull type (after Romer, 1966). *E, Batrachosuchus* (after Romer, 1966). *F,
Trematosaurus* (after Romer, 1966).

jaws and sharp conical teeth indicate that they were predators, and the large palate teeth suggest that they fed on relatively large prey—probably fishes and possibly on each other. Some of the smallest labyrinthodonts were relatively lightly built, proportioned like modern lizards, and, like lizards, they probably fed mainly on terrestrial invertebrates (insects and the like). During the first 100 million years of their history, some lines of labyrinthodonts became better adapted for land life. They developed stronger limbs and vertebral columns and probably were completely terrestrial predators. Only one group (the diadectids) appears to have become herbivorous. We make this inference from their large bodies (up to 10 feet in length), their very small heads, and small, blunt, transversely widened teeth. Flattened teeth are characteristic of living herbivores from elephants to fish. The shape functions to increase the surface area available for grinding down tough plant material. The length of the jaws is highly correlated with the length of the head. As we learned in chapter 8, the length of the jaw is the out-lever in the system producing out-force at the tooth row. Shortening the out-lever is one biomechanical way to increase the out-force. Large forces are needed to break down tough plant material. It is surprising that only one group of amphibians became herbivorous, since there was an abundance of vegetation on land.

As some lines of labyrinthodonts showed strengthened limbs and vertebral columns as adaptations for life on land, other lines reduced the size and strength of their limbs; developed long, flexible bodies with weak vertebral columns; retained their larval lateral line system into adulthood (fossils show us the presence of grooves on their dermal skull bones); and, in some groups, developed high, laterally flattened tails (Eogyrinus, fig. 10.2). Clearly, these labyrinthodonts had become secondarily adapted to a fully aquatic life. A common evolutionary trend among this group was the development of increasingly flattened heads, suggesting that they spent much time lying on the bottom. Other lines of aquatic labyrinthodonts evolved very long and narrow snouts, like those of modern fish-eating crocodiles. Some labyrinthodonts show evidence of external gills in their juvenile stages, like those of larval salamanders, which indicates that they reproduced like modern amphibians by laying eggs in the water, followed by an aquatic larval stage.

The second grouping of early amphibians, the lepospondyls, were mostly very small, less than 1 foot long, with relatively long, slender bodies and short weak legs (fig. 10.3). The enamel of their small teeth lacked the labyrinthine structure seen in their larger relatives, and their

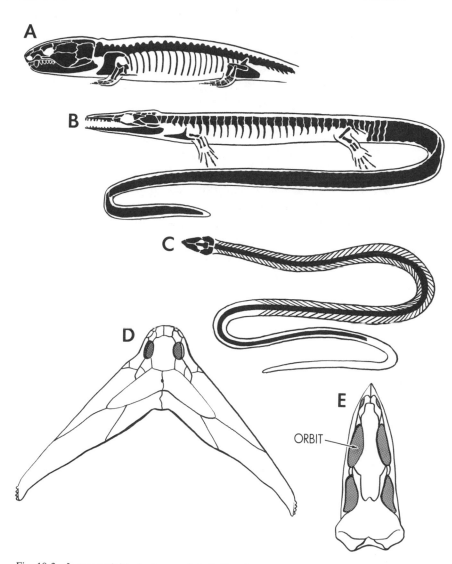

Fig. 10.3   Lepospondyl body shape and skull diversity. *A, Pantylus,* about 6 inches long (after Carroll, 1968). *B, Sauropleura,* about 1 foot long (after Milner, 1980). *C, Ophiderpeton,* about 2.5 feet long (after Romer, 1966). *D, Phlegethontia* (after Beerbower, 1963). *E, Diploceraspis* (after Romer, 1966).

skulls lacked a notch for the eardrum. In most groups of lepospondyls, the notochord was surrounded by a spool-shaped centrum beneath each neural arch.

The small body size and small sharp teeth of the lepospondyls suggest that they fed mainly on small invertebrates. Many lines among the group called nectrideans evolved even more elongate bodies, weaker limbs, and a less well ossified vertebral column and appear to have become secondarily completely aquatic. Like the aquatic labyrinthodonts, they often had lateral line grooves on the skulls and laterally flattened tails, and some fossils show evidence of external gills. One of the most unusual of the aquatic lepospondyl groups, the diplocaulids, developed relatively large, flattened, triangular skulls, with the dorsal eyes and nostrils of bottom dwellers but with grotesquely elongated skull corners (fig. 10.3). We have no modern analogues of these latter types for form-function extrapolation, and optimal design analysis provides few clues to their way of life. A third group of lepospondyls, the aistopods, lost their limbs and girdles and developed a very long body with a strongly ossified, pointed skull (fig. 10.3). We think that they were probably aquatic, but they could have lived in ground litter much like modern snakes and short-limbed or limbless lizards.

After almost 100 million years of success, most lines of amphibians became extinct. Some possibly could not compete with early reptiles, and others perhaps were casualties of a massive wave (or waves) of extinctions that occurred between 250 and 225 million years ago. A few groups of labyrinthodonts survived the great die-out and gave rise to a small second evolutionary radiation. This time most were large aquatic forms, including some with skulls up to 3 feet in length! This radiation lasted only about 30 million years, for by 190 million years ago the last of the great labyrinthodonts were extinct, perhaps outcompeted by the burgeoning groups of aquatic reptiles.

Modern amphibians first appear in the fossil record around 200 million years ago (frogs) and 140 million years ago (salamanders), looking very much like their living descendants. Salamanders (order Caudata, or tailed ones), retain the primitive body proportions seen in small labyrinthodonts and lepospondyls (fig. 10.4) but have lost the scales that covered the early amphibians. Terrestrial salamanders use both lungs and skin for respiration; aquatic salamanders use external gills. There are around 325 living species, ranging in length from a few inches to 4 feet, although most are about 4–6 inches in length. They are often found in cool, moist habitats, living under stones, logs, and leaf litter. One group of salamanders, however, has undergone a radiation in the new world tropics, living in hot, moist habitats. Salamanders feed on small invertebrates. Their most unusual morphological modification has been the development in some groups of an elaborate tongue and hyoid

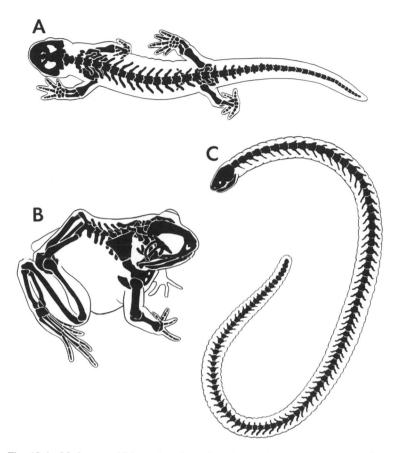

Fig. 10.4 Modern amphibians. *A,* salamander, about 5 inches long (after Schaeffer, 1941). *B,* frog, 3 inches long (after Gans, 1974). *C,* caecilian, about 16 inches long.

apparatus that is capable of shooting the tongue out almost half a body length to catch insects.

Frogs and toads (order Anura, or without a tail), are much transformed from primitive amphibian proportions. Their heads are relatively large, their trunks are very short, they have no tails as adults, their hind limbs are relatively enlarged, and their pelvic girdle is lengthened (fig. 10.4). These morphological specializations are related to their unique jumping locomotion. Like salamanders, anurans are scaleless and use both lung and skin respiration. Most frogs have an aquatic larval or tadpole stage, and, unlike the larvae of salamanders, tadpoles are constructed very differently from adults. Most tadpoles have large mouths and guts, are filter feeders, and have a small tail for propulsion. They undergo a dramatic change to the adult anuran body plan at metamorphosis and change their diet to insects and other small invertebrates. There are about 3,000 species of frogs and toads alive today.

Most frogs and salamanders reproduce in water and have aquatic larvae. Frogs usually have external fertilization while salamanders have internal fertilization. But modifications of the general pattern have evolved independently a number of times in both groups. For example, internal fertilization is now known to occur in about a half dozen different frogs. A more common modification of the basic reproductive pattern is called direct development, which occurs in some salamanders and frogs. In these cases, the larvae do not go through an aquatic stage but instead develop directly in the egg into small juveniles. In direct development, the eggs are laid on land in moist areas under rocks or in moss. Direct development frees the amphibians from having to return to water for reproduction, but they are still restricted to moist habitats on land.

The last order of modern amphibians, the Gymnophiona, is a little-known group of about 160 species restricted to the tropics. These worm-like animals are commonly called caecilians (fig. 10.4). Some species have small, dermal scales hidden under the skin, and many species are blind or have reduced eyes hidden under the skull bones. Most members of this group are terrestrial burrowers, but there are aquatic forms as well. Among their unusual features are the loss of limbs and girdles, the development of a pair of chemo-sensitive tentacles on their snouts, and the development of the ability of over half the species to give birth to live young rather than lay eggs. Fertilization in these forms is internal, and the young derive nutrition from a secretion that the mother produces before birth. Like salamanders and frogs, caecilians are relatively small (up to 2 feet) and feed on invertebrates.

Early Reptiles

CHAPTER 11

The labyrinthodont and lepospondyl amphibians, like most modern frogs and salamanders, probably returned to the water to lay their eggs or at least were restricted to very moist habitats for reproduction. About 310 million years ago (around December 7), 50 million years after the first appearance of amphibians, a major breakthrough in vertebrate evolution occurred: a group of small labyrinthodont amphibians evolved the ability to lay their eggs on dry land, away from water. Like the direct-developing amphibians, these animals lost their larval stage, and the young hatched from the egg looking like miniature adults. This mode of reproduction required the evolution of internal fertilization and involved the evolution of several new structures in the egg. The structures, called extraembryonic membranes, are the amnion, which encloses the embryo in a fluid-filled sac; the chorion, which lines the egg shell and allows for gas exchange; and the allantois, which provides a sac in which nitrogenous metabolic waste products are stored away from the embryo. This type of egg is called an amniote egg, and we infer that it was present in the earliest reptiles because all of their living descendants (modern reptiles, birds, and mammals), some of which diverged early from each other, have an amniote egg (fig. 11.1).

While the amniote egg was the most important feature that distinguished the earliest reptiles from their amphibian ancestors, they also differed in some relatively minor skeletal features that allow us to recognize them as reptiles from their fossilized bones. Early reptile skulls tend to be relatively smaller, narrower, and deeper than those of most labyrinthodonts. Differences in the structure of the back of the skull roof and braincase have suggested to some paleontologists that the jaw muscles that attached there were shifted so that there was improved mechanical efficiency for maximizing force at all stages of jaw closure. There was no notch at the back of the skull for an eardrum, and the stapes was a massive bone that braced the cheek bones against the brain-

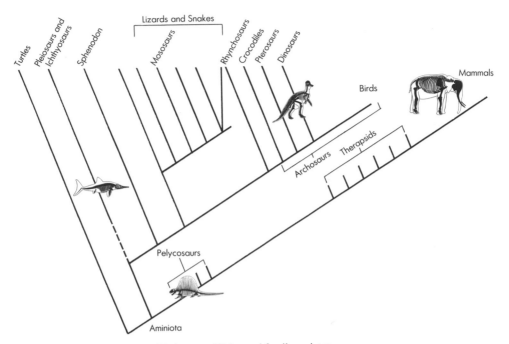

Fig. 11.1　Phylogeny of living and fossil amniotes.

case. Eardrums and the use of the stapes for sound transmission evolved in later reptiles, independently of their evolution in amphibians. The wrist and ankle regions in early reptiles appear to have been shortened and strengthened by fusion and by a reduction in the number of separate bones. Generally, the early reptile bones were more slender than those of amphibians. In addition, early in reptile evolution the pelvic girdle expanded its contact with the vertebral column from one to two sacral vertebrae. The expanded connection may have been correlated with an increase in the amount of force the hindlimb generated, and/or it may have functioned to strengthen the link between the girdle and vertebral column to transfer more of the propulsive thrust of the hindlimb to the body.

Another feature that was probably present in early reptiles was a new type of body covering. In place of the heavy bony dermal scales of their labyrinthodont ancestors, reptiles developed a light but tough, flexible, horny scale, grown in the epidermis and composed largely of keratin. In addition, reptiles retained the ability to grow bony plates in the dermis layer of the skin, and in many groups of reptiles dermal bone underlies epidermal horny scales. It should be noted that modern teleost fish achieved the same functional end—a light, flexible body covering—by greatly thinning their dermal bony scales. The functional significance of having reptilian horny scales versus teleost thin, bony ones is not obvious.

The earliest reptiles were small and superficially lizard-like in appearance, and like modern lizards they probably fed on insects and other small invertebrates. Shortly after their first appearance, they underwent a modest evolutionary radiation, amid the radiations of the labyrinthodont and lepospondyl amphibians (fig. 11.2). Most groups remained relatively small, ranging up to about a foot or two in body length, with

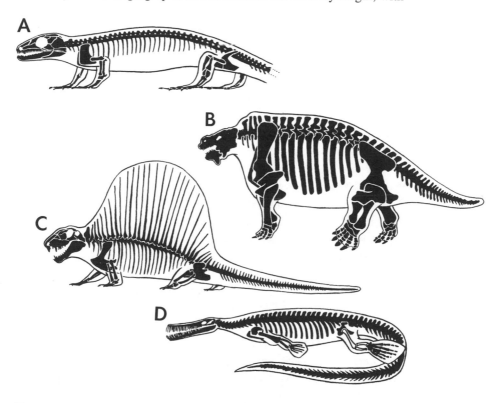

Fig. 11.2 Early reptile body shape diversity. A, *Paleothyrus*, about 6 inches in body length, representing primitive proportions (after Carroll, 1969). B, *Scutosaurus*, an herbivore about 10 feet long (after Gregory, 1951). C, *Dimetrodon*, a carnivore about 8 feet long (after Romer and Price, 1940). D, *Mesosaurus*, an aquatic reptile about 16 inches long (after Romer, 1966).

small, sharp, conical teeth, suggesting that they retained the insectivorous feeding habits of their ancestors. A few lines (captorhinids, procolophonids) evolved expanded, low, blunt teeth that exhibit heavy wear, suggesting that they fed on plants or hard-shelled invertebrates. The pareiasaurs grew to the relatively large size of about 8 feet in length. They had relatively small heads, often fringed with rugosites and blunt spikes; relatively small, blunt teeth; and massive, ponderous bodies, and they were among the largest land vertebrates of their time.

One of the most successful group of early reptiles, in terms of diversity, were the pelycosaurs. They are distinguished from other reptiles by the presence of a small opening in a characteristic position in the side of the skull behind the orbit. Such openings developed in unstressed regions of the skulls of other groups of reptiles as well, and we think initially that they provided firmer attachment areas for jaw muscles. Later, as they enlarged, the openings allowed increased space for larger jaw muscles. Differences in the locations and numbers of these temporal region openings have been used to infer phylogenetic relationships among the major groups of reptiles.

From relatively small, insectivorous ancestors, pelycosaurs diversified into small and large carnivorous types and two groups of herbivores. Some of the carnivorous pelycosaurs had long, narrow jaws lined with a large number of small, conical teeth, a form that suggests that they were fish eaters. Others, like the large *Dimetrodon* (as long as 8 feet), developed a reduced number of enlarged, slightly compressed, sharp teeth, with one or two pairs of large fangs toward the front of the tooth row, followed by smaller teeth to the rear. *Dimetrodon* was the largest predator of its time, and its enlarged anterior teeth suggest that it fed on large terrestrial vertebrates. It was also unusual in having extraordinarily long neural spines (fig. 11.2), which in life supported a web of highly vascularized (i.e., with many blood vessels) skin.

Living reptiles are ectothermic; they use external sources of heat (namely, the sun) to regulate body temperature. Living birds and mammals, on the other hand, are endothermic, and they use internally generated heat to regulate body temperature. Endotherms maintain a constant high body temperature and metabolic rate. Ectotherms generally have lower metabolic rates, and their body temperature fluctuates with external temperatures, at least to some extent. Studies of lizard behavior show that to raise their body temperature they bask in the sun, orienting their bodies so that the maximum surface area for heat absorption is perpendicular to the source of heat. When minimizing heat uptake, they change the orientation of their bodies to a position parallel to the heat source, therefore presenting a minimal surface area to the sun. The basking behavior in living reptiles suggests that the sail of *Dimetrodon* was an adaptation that improved its ability to regulate body temperature. It may have provided a large surface area (1) to pick up heat and warm the body when it was cold and (2) from which to radiate heat and cool the body when it was too warm.

Another possibility is that the sail functioned as a display organ. The males of some species of living lizards have flaps of skin under their throats that can expand and that they show to other males. This display behavior is important in territorial defense, mate choice, and species recognition. Each species with such a flap has a unique color pattern on it. Temperature regulation and display are not mutually exclusive, and the sail of *Dimetrodon* may have functioned in one or both capacities. There is no way, unfortunately, that we can ever determine from the fossil record itself which problem was paramount in the original evolution of the sail. Biomechanical analysis tells us only that the sail is a good solution, in this case, for more than one problem.

The herbivorous pelycosaurs are distinguished by very small heads and small, paddle-shaped teeth (fig. 11.3). They grew to relatively large

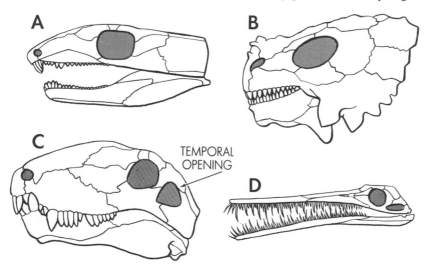

Fig. 11.3   Early reptile skull shape diversity. *A, Eocaptorhinus,* representing the primitive type (after Heaton and Reisz, 1980). *B, Scutosaurus* (after Heaton, 1980). *C, Dimetrodon* (after Heaton, 1980). *D, Mesosaurus* (after Romer, 1966).

size (6–8 feet in length), and one genus, *Edaphosaurus,* evolved a neural spine "sail" like that of *Dimetrodon.*

Of all the groups that emerged in the early radiation of reptiles, only
two appear to have been fully aquatic. One of these, the mesosaurs, was
relatively small (up to a few feet in length) and had the aquatic adaptations of enlarged, paddle-like hind feet; long, laterally compressed tails;
and swollen, heavy ribs (which may have acted as ballast; see fig.
11.2). Mesosaurs had very long, narrow jaws lined with a large number
of long, needle-like teeth, which suggests that they fed on very small
fish or aquatic invertebrates. The evolution of so few aquatic forms in
the early reptile radiation is in strong contrast to the amphibian radiation, which produced a great number of secondarily aquatic lineages.
Perhaps the fact that there were many different kinds of amphibians
already present in shallow waters when reptiles first appeared acted as a
block that prevented the reptiles from a major invasion of that habitat.

About 250 million years ago (around December 11), many lines of
early reptiles, as well as many labyrinthodonts and all lepospondyl amphibians, became extinct. This extinction took most of the pelycosaurs,
except for a lineage of small carnivores that survived extinction and
rapidly diversified into an evolutionary radiation of forms called therapsids, or mammal-like reptiles. The therapsids flourished for almost 50
million years and are represented by a great variety of insectivorous,
carnivorous, and herbivorous types. One of the early groups, the dinocephalians (terrible heads), included several genera of large herbivores
8–10 feet in length, with unusual cranial features. Their greatly thickened skull roofs and reinforced braincases suggest, from optimal design
analysis, that dinocephalians engaged in head butting (fig. 11.4). Given
the ponderous build of these large reptiles, it seems unlikely that they
engaged in high-speed collisions, but, even if the behavior were more
like shoving, it suggests a complex social structure, presumably involv-
ing intraspecific competition for mates or territory.

Another group of therapsids with unusual cranial anatomy were the
dicynodonts, small to large herbivores 1–10 feet in length, which modified their jaw apparatus in a manner radical for tetrapods of their time.
They reduced or lost all of their teeth except for a pair of large tusks
(hence the name dicynodont, or two doglike teeth), developed turtle-
like beaks that in life were probably covered with a horny sheath, and
had greatly enlarged temporal openings (in the skull wall behind the
orbit) that accommodated a large mass of jaw musculature (fig. 11.4).

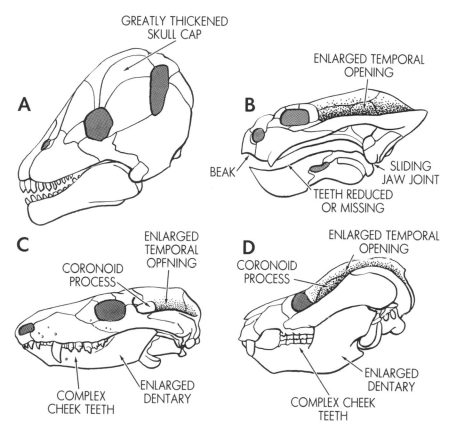

A   GREATLY THICKENED
     SKULL CAP

B   ENLARGED TEMPORAL
     OPENING
     BEAK
     SLIDING
     JAW JOINT
     TEETH REDUCED
     OR MISSING

C   ENLARGED
     TEMPORAL
     OPENING
     CORONOID
     PROCESS
     COMPLEX
     CHEEK TEETH
     ENLARGED
     DENTARY

D   ENLARGED TEMPORAL
     OPENING
     CORONOID
     PROCESS
     ENLARGED
     DENTARY
     COMPLEX CHEEK
     TEETH

Fig. 11.4 Therapsid skull diversity. *A, Moschops,* a dinocephalian head butter (after Barghusen, 1975). *B, Dicynodon,* a specialized herbivore (after Crompton, 1972). *C, Thrinaxodon,* an insectivore (after Romer, 1966). *D, Bienotherium,* an advanced herbivore (after Kemp, 1982).

Reconstruction and functional analysis of the jaw apparatus suggests that the dicynodonts bit off pieces of vegetation with their beaks and then sheared or sliced it between tooth remnants or horny plates located on the palate, using a posterior movement of their lower jaws. This is indeed a radical departure from what is inferred to have been the vertical crushing action of the jaws of other early reptilian herbivores.

The early carnivorous therapsids were mostly medium-sized to large, ungainly, sprawling creatures like their pelycosaur predecessors. However, in the latter part of their evolutionary history, from about 225 to 200 million years ago, they are represented by small- to medium-sized forms called cynodonts (dog toothed) that display changes in their feeding and locomotor systems (fig. 11.4). The lower jaw came to be composed largely of a single bone (the dentary), and the remaining original bones were reduced to tiny remnants at the back of the jaw. The tooth row became more differentiated, with incisors in front, followed by canines and cheek teeth; in later forms, the cheek teeth developed complex crowns. In the upper jaw, a horizontal shelf of bone grew inward to form a secondary palate that separated an air passage above from a food passage below.

Recent experimental work shows that a secondary palate reinforces the skull against forces generated by the jaw muscles during jaw closing or feeding. A secondary palate represents a potential solution for two different problems—one of separating air and food passages and the other of skull support against forces generated during feeding. Optimal design analysis cannot distinguish which problem was originally responsible for the evolution of a secondary palate, and secondary palates have evolved independently a number of different times in vertebrates. There may have been different selective forces, or problems to solve, that shaped its evolution in various groups. Correlation analysis may, in some situations, help to sort out which problem was involved in any particular evolutionary acquisition of the secondary palate. Separating air and food passages is important for aquatic animals and for animals that spend a fair amount of time chewing food in the mouth. In those forms in which a secondary palate evolved in response to the problem of living in an aquatic environment, we might look for correlated structures related to aquatic locomotion. In response to the problem of skull loading during feeding, we might expect to see correlated changes in skull morphology such as an increase in the area of the jaw muscle attachment. Nonetheless, it should be clear through this and other examples that biologists often have no way to reconstruct the selective forces that were responsible for the origin of a new morphological feature.

Finally, the temporal opening at the back of the dicynodont skull became enormously enlarged, and from the attachment scars we conclude that the jaw musculature had become more differentiated. This cluster of changes in the feeding apparatus suggests the development of more powerful jaw muscles that were more mechanically efficient and that gave more precise control of jaw movement, stronger jaws, and more thorough chewing of food in the course of cynodont evolution.

In the locomotor system, changes in the shape of limb joints suggest more upright, less sprawling posture. The lower elements of the pectoral girdle were reduced in size, and the upper portion (ilium) of the pelvic girdle was expanded. These changes suggest a reduction in the extent of the ventral girdle musculature (used to raised the body off the ground in the sprawling position), and an increase in the size of the hind limb musculature involved directly in propulsion. These changes in the locomotor system suggest that later cynodonts may have moved at greater speeds compared with what we infer for their pelycosaur and early therapsid ancestors. Coupled with the changes in their feeding system, they suggest a higher level of activity.

The early cynodonts were medium-sized carnivores, but most of the later cynodonts, judging from their small size, were probably insectivorous. Some of the later cynodonts developed large cheek teeth with relatively flat, complex crowns that suggest an herbivorous diet. It is from one line of small insectivorous cynodonts that mammals are thought to have evolved, and that story will be continued in chapter 16.

By about 200 million years ago (around December 17), the great therapsid radiations were largely over. Reptile land faunas came to be dominated by other groups that had evolved in a great radiation from what had been insignificant lizard-like survivors of the extinctions that had decimated the land tetrapod faunas 50 million years earlier. We will turn now to these later reptile radiations, which produced some of the most spectacular animals ever to walk the earth.

The Great Reptile Land Radiations

As the therapsids declined, between about 225 and 200 million years ago (December 17–19), other groups of reptiles radiated to replace them. The most important of these, in terms of eventual diversity and abundance, were the archosaurs (ruling reptiles). The earliest archosaurs were relatively small, lightly built, and generally lizard-like in appearance. Their limbs and girdles were little changed from the primitive reptile condition, and their main distinguishing features were in the skull. There were two small temporal openings behind the orbit (the diapsid condition), one in front of the orbit and one in the lower jaw (*Euparkeria;* see fig. 12.4 below). In addition, the teeth were set in sockets rather than attached along the edge of the jaw. The openings in the skull and jaw suggest some reorganization in, or at least more efficient attachment of, the jaw musculature, but a more precise functional significance of those openings in their particular positions remains to be worked out. The small body size and type of dentition (large number of small, sharp teeth) suggest that the earliest archosaurs fed on insects and small vertebrates. However, they rapidly diversified into a great variety of small to gigantic terrestrial herbivores and carnivores, the most spectacular of which are the dinosaurs, but which included some lesser-known, interesting forms as well (fig. 12.1).

Among the early experiments in carnivory was a group of archosaurs called phytosaurs, which looked superficially like modern crocodiles. They were large animals, as long as 10–15 feet, with relatively long bodies; long, laterally flattened tails; and short, sturdy legs. The laterally flattened tail suggests that they were aquatic, like modern crocodiles, and the sturdy limbs indicate that they also walked around on land. Phytosaurs had very long, narrow jaws lined with a large number of similar-sized, sharp, conical teeth, from which we infer that, like modern crocodiles, phytosaurs were primarily fish eaters. An unusual modification of the skull is the anterior elongation of one of the skull

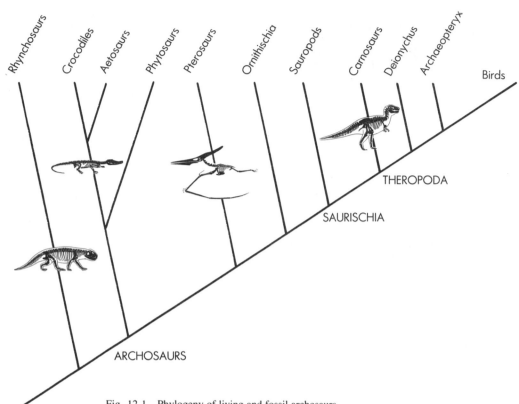

Fig. 12.1   Phylogeny of living and fossil archosaurs.

bones (premaxilla). As a result, the external nostrils are in a position just in front of the orbits (fig. 12.2), quite far back from the front of the snout. The nostrils are also elevated. The roof of the mouth in these forms is recessed into the floor of the skull; this probably had the effect of isolating the air passageway from the mouth chamber, thus allowing breathing to continue while the mouth was full of food and water.

Contemporaneous with the phytosaurs were a group of similar-sized and similarly proportioned archosaurs, but this second group had smaller heads, shorter jaws, and relatively small, conical or blunt teeth. These were the aetosaurs, notable for evolving an extensive covering of dermal bones (fig. 12.2), which were probably capped with a covering of tough, horny epidermal scales, like armor over most of the body. Their small heads and blunt dentitions suggest that they were herbivores.

The phytosaurs and aetosaurs did not last long, geologically speak-

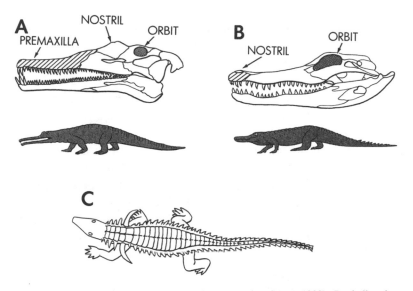

Fig. 12.2   A, skull and body outline of a phytosaur (after Camp, 1930). B, skull and body outline of an alligator (after Colbert, 1980). C, body shape of an aetosaur. Animals are 10–15 feet long.

ing. They became extinct after only about 10 million years, around the time a third group of reptiles, the crocodilians, arose from early archosaur stock. This third group was more successful in terms of longevity, for more than 200 million years later its conservative descendants—the modern crocodiles, alligators, and gavials—are still with us today.

Crocodilians have relatively long bodies, with long, laterally flattened tails; short, sturdy limbs; and long-jawed, flattened skulls. Their limbs and girdles are somewhat modified from the primitive reptilian condition. In water, the laterally compressed tail is the main propulsive organ. On land, crocodilians are capable of both fast and slow locomotion. Modern crocodiles have even been seen to bound much like a running dog. During slow walking, they hold their limbs in a sprawling posture. When running, the limbs are placed more directly under the body. This shift in the position of the limb results in a longer stride length (see detailed discussions below) and faster locomotion. The main modification of the skull of crocodilians has been the development of a secondary palate, analogous to that of late therapsids and mammals, that separates the air passageway above from the mouth chamber below. This allows crocodilians, which have terminal, not posteriorly displaced nostrils, to breathe while they have water in their mouths (fig. 12.2).

Skull shape varies from extremely narrow snouted in the purely fish-eating gavials, to broader snouted in the alligators and caimans, which feed on turtles and other vertebrates as well as on fish.

There are about twenty-five species of living crocodilians, ranging from small caimans a few feet long to large crocodiles as long as 20 feet. Almost all are freshwater forms, although some of the largest crocodiles inhabit brackish estuaries and river mouths and go far out to sea. Between about 200 and 135 million years ago, there were giantic marine crocodilians as long as 40–50 feet. These animals, called thallatosuchians, had elongate webbed digits making paddle-like hands and feet and a fishlike tail in which the vertebral column turned down into the lower lobe. Between 50 and 15 million years ago, there were two lines of more terrestrial crocodilians, with limb modifications that suggest a habitually erect posture and deep skulls with bladelike teeth.

The phytosaurs, aetosaurs, and crocodilians constitute a minor branch of the archosaurs. Far more spectacular, in terms of body modifications and evolutionary radiations, was the line that produced the dinosaurs. Although dinosaurs included the largest animals ever to walk on Earth, their ancestors were small (about 4 feet long), lightly built, bipedal (two feet on ground) archosaurs (fig. 12.3). The main modifications from the primitive archosaur condition were in the locomotor system. Reflecting the shift to bipedal locomotion, the forelimbs were relatively short and slender, and the pectoral girdle, no longer involved in weight support, was greatly reduced in size. The hind limbs were relatively long and were held under the body, and the ilium (upper portion of the pelvic girdle) was greatly enlarged, providing expanded contact with the vertebral column for weight transmission and a greater area for attachment of locomotor muscles.

The body was held erect. The primitively platelike lower portion of the pelvic girdle was transformed to become a divergent, rodlike pubis and ischium, reflecting a reduction in the massive ventral support musculature needed with a more primitive sprawling posture. The limbs moved under the body and could directly support more of the body weight in a column-like fashion. The ventral support musculature was reoriented and became incorporated into larger protractor and retractor muscles of the limb. The ankle region was modified for more efficient weight transmission and for increased fore-aft rotation at the joint. The animals appear to have walked up on their fingers and toes rather than flat-footed.

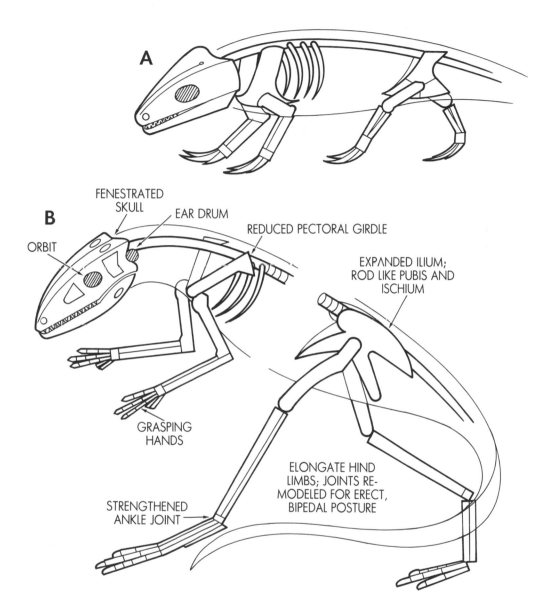

A

B

FENESTRATED
SKULL

EAR DRUM

REDUCED PECTORAL GIRDLE

ORBIT

EXPANDED ILIUM;
ROD LIKE PUBIS AND
ISCHIUM

GRASPING
HANDS

ELONGATE HIND
LIMBS; JOINTS RE-
MODELED FOR ERECT,
BIPEDAL POSTURE

STRENGTHENED
ANKLE JOINT

Fig. 12.3   Advanced archosaur body plan. *A, Eocaptorhinus,* representing a primitive reptile morphology. *B,* an advanced archosaur, modeled after *Lagosuchus.*

The shift to an erect posture, vertically placed limbs, fore-aft rotation of the joint, and the use of the tips of the toes to make contact with the ground are related to shifts in locomotor pattern. As we noted earlier, increased speed of movement can be accomplished two ways, that is, by lengthening each step and/or by reducing the time that it takes to make a single step. The shift in the position of the dinosaur limbs meant that more of the leg was involved in stepping as opposed to supporting the body, as is the case with animals that have a sprawling posture. By walking on the tips of the toes instead of flat-footed, the length of the foot was added to limb length. Both these morphological shifts increase step length. The increase in the area of the pelvic girdle indicates an enlarged musculature to move the limb. Not only was this muscle mass larger, but it also inserted proximally on the elongated leg. This type of insertion, close to the joint, increases the velocity of limb movement or frequency of step cycle. A large portion of the energy used in running is spent accelerating the limbs back and forth. The resistance to movement of any object is called its moment of inertia. The higher the moment of inertia, the more energy it takes to move the object. Moment of inertia is proportional to the square of the distance from the center of mass of an object to its pivot point. Now think of a limb. The pivot point for a leg is the hip joint. If the distal part of the limb has heavy musculature, the center of mass will be a longer distance from the hip joint. By locating the musculature of the limb close to the hip joint, the center of mass of the limb is moved higher up on the leg, much closer to the pivot point. This results in a lower moment of inertia and increases the speed of leg movement for a given muscle contraction or energy expenditure. The reorganization of the locomotor system, then, seems to have been related to increasing speeds of movement.

Two specialized groups of dinosaurs evolved from this ancestral form, about 200 million years ago (about December 17). These were the saurischians (reptile hipped), and the ornithischians (bird hipped). The two groups are distinguished by their different pelvic girdle structure. The saurischians retained a pelvic girdle similar to that of the ancestral form, with divergent, rodlike pubis and ischium. In the ornithischian line, the pubis rotated backward to brace the ischium, and the anterior portion of the ilium extended forward to provide an attachment area for protracting muscles. In advanced ornithischians, the anterior portion of the pubis was expanded forward to supplement the anterior ilium for protractor muscle attachment area. The saurischian and ornithischian pelvic girdles appear functionally equivalent. Both types provide attachment sites for the modified protractor and retractor musculature correlated with the shift from sprawling to erect posture. As such, they are good examples of alternative solutions to similar problems.

The saurischian dinosaurs include both carnivores and herbivores (fig. 12.4). The carnivores, called theropods (not to be confused with the mammal-like reptile therapsids), remained bipedal and ranged from animals 3–4 feet in length to gigantic specimens as long as 40 feet. The small carnivores were very lightly built and had long, flexible necks and grasping hands similar to those of their bipedal archosaur ancestors. They had small, laterally compressed, sharp teeth replaced by a horny beak in some later forms and probably fed on insects as well as small vertebrates. The larger forms, like the famous *Tyrannosaurus* (fig. 12.5), grew as long as 40 feet and stood almost 20 feet high and were the largest land carnivores that ever lived. They had relatively large heads, with large, bladelike teeth; small forelimbs; relatively large, powerful hindlimbs; and a long, heavy tail. The large tail would have served as a counterweight to balance the horizontally held head and trunk during bipedal locomotion and as a prop (a "third leg") during sitting. Forelimbs were progressively reduced in size in later large theropods, and in the latest, gigantic species they were so small that prey capture and killing must have been done with jaws and hindlimbs alone. A recently discovered variant on these gigantic predators is the medium-sized *Deinonychus,* which was about 10 feet long. This theropod had secondarily enlarged forelimbs and hands and large claws on the hands, which were used along with the gigantic claws on the hind feet for subduing and killing prey. *Deinonychus* is an important dinosaur in the story of vertebrates because birds evolved from this group.

The herbivorous saurischians, called sauropods, evolved early to a large size (10–20 feet in length) and then to truly gigantic lengths as great as 60–80 feet (fig. 12.5). With the increase in size, they reverted to a secondarily quadrupedal posture, but maintained their fore and hindlimbs directly under the body rather than positioned laterally in a sprawling stance.

Muscular contraction uses energy. When the limbs are laterally placed, muscular contraction is required to support the body off the ground. When the limbs are directly under the body, they can act as columns of support and muscles play a smaller role in body support. For this reason, posture with limbs placed under the body is more efficient

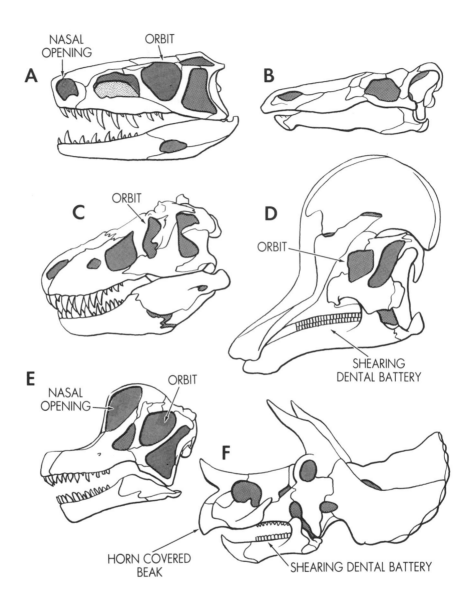

Fig. 12.4   Dinosaur skull diversity. *A, Euparkeria,* a primitive archosaur (after Ewer, 1965). *B, Stegosaurus,* a stegosaurian (after Romer, 1966). *C, Tyrannosaurus,* a carnivorous saurischian (after Gregory, 1951). *D, Corythosaurus,* an ornithopod (after Ostram, 1961). *E, Brachiosaurus,* an herbivorous saurischian (after Gregory, 1951). *F, Triceratops,* a ceratopsian (after Romer, 1966).

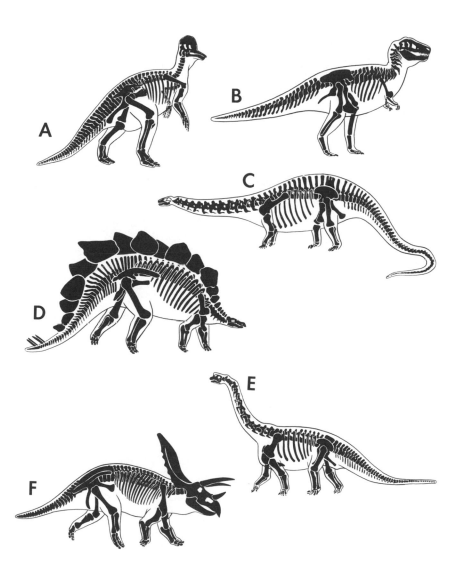

Fig. 12.5 Dinosaur body shapes. *A, Corythosaurus,* a hadrosaur, about 30 feet long (after Charig, 1979; Colbert, 1980). *B, Tyrranosaurus,* a theropod carnisaur, about 40 feet long (after Gregory, 1951). *C, Brontosaurus,* a sauropod, about 80 feet long (after Gilmore, 1936). *D, Stegosaurus,* about 18 feet long (after Romer, 1966). *E, Camarosaurus,* a sauropod, about 19 feet long (after Gilmore, 1925; Bakker, 1980). *F, Torosaurus,* a ceratopsian, about 25 feet long (after Young, 1962; Bakker, 1980).

for weight support than a sprawling posture. Additionally, muscles alone could not support a body the size of a sauropod. We know this from studies of large, living mammals. Elephants, rhinoceroses, and hippopotami have all shifted their limbs under their massive bodies to act as columns of support, and these animals are not nearly as large as some of the sauropods.

Sauropods had tiny heads, small and weak teeth, very long necks and tails, and relatively short trunks. The larger sauropods were the largest terrestrial animals that ever evolved, and earlier reconstructions figured them as ponderous swamp dwellers, too heavy to walk about on land. However, recent functional analyses of limb bones and joints and comparisons with large modern terrestrial animals, like elephants, suggest that the sauropods could indeed walk on land and that they probably held their long necks erect and browsed off the tops of tall trees (fig. 12.5). Their mouths are too small and their teeth too few and blunt to have been the main processors of enough vegetation to support such enormous bodies, and we know from some fortunate finds that the sauropods had a gizzard grinding mill in the stomach, where swallowed stones ground up vegetation at the beginning of digestion. Many sauropods had greatly enlarged nasal openings, located just in front of the braincase high on the skull. A recent ingenious suggestion for their functional significance is that they were part of a system to cool blood flowing to the brain. If blood is too warm it can damage the brain cells.

The second major grouping of dinosaurs, the ornithischians, were all herbivores and included four very distinct groups: ornithopods, stegosaurs, ankylosaurs, and ceratopsians. All grew to quite a large size; most species were between 10 and 30 feet long as adults. All four groups replaced their front teeth with horny pads or beaks, and the ornithopods and ceratopsians developed impressive grinding and shearing dental batteries from the posterior dentition (fig. 12.4). The ornithopods retained bipedal locomotor capabilities, while the other three groups were secondarily quadrupedal, with relatively long hindlimbs, plus the ornithischian girdle type, as reminders of their bipedal ancestry.

There were three notable morphological developments among the ornithopods (bird footed). First, some species in a subgroup called pachycephalosaurs (thick-headed reptiles) grew enormously thickened skull caps, 8 inches or more of solid bone, over their braincases and modified the skulls to optimize absorption of impact loading. This phenomenon is similar to what had evolved among the dinocephalian therapsids many millions of years earlier, although it is more extreme, and we may infer a similar behavioral complex: competition for status (or territory or mates) in a dominance hierarchy. Further, the pachycephalosaurs were bipeds and capable of running at high speeds, so we may infer that their skulls withstood high-speed collisions, rather than just shoving, in their head-butting contests.

Second, among the group of ornithopods known as hadrosaurs, or duck-billed dinosaurs (named for their broad, flat beaks), we find an extraordinary diversity of bony crests developed on top of the heads (see fig. 12.4). Formed by the backward elongation of anterior skull bones (premaxillaries and nasals), some crests were solid bone, but other crests contained within them a greatly elongated and diverted nasal passage. By analogy with modern lizards that have various crests or dewlaps around their heads, it has been suggested that the hadrosaur crests served as visual signals for species recognition and/or intraspecific sexual selection. A most ingenious suggestion for the functional significance of the hollow crests is that they may have acted as resonating chambers for the production of loud vocal signals.

The third unusual morphological development among ornithopods was the modification of the skull and dentition for grinding tough vegetation. The hadrosaurs had dental batteries composed of densely packed rows of literally hundreds of tiny teeth that appear, from biomechanical design analysis and reconstruction of jaw movements, to have functioned for shearing and grinding tough vegetation. Finally, we should not leave the hadrosaurs without mentioning that, although they had relatively large, powerfully built hindlimbs, suggestive of terrestrial bipedal locomotion, they also had long, laterally compressed tails, from which we infer that they could engage in aquatic locomotion as well. As confirmation of this hypothesis, a few specimens have been found that preserve impressions of webbing on hands and feet, indicating that hadrosaurs were amphibious dinosaurs, with adaptations to enhance both terrestrial and aquatic locomotion.

The second group of ornithischians, the stegosaurs (plated reptiles), were long-bodied, small-headed reptiles, most notable for having two rows of erect, diamond-shaped plates extending back along their vertebral columns and two pairs of large spikes at the tips of their tails (fig. 12.5). For quite a long time biologists thought these plates served as armored protection for the vertebral column. The plates, however, were relatively thin and covered with grooves from blood vessels, which is not what one would expect of armor. Recent wind tunnel experiments with models of stegosaurs suggest that the plates are optimally designed and placed to serve as heat radiators, from which we may infer that they functioned for thermoregulation. The inferred presence of extensive blood vessels is further support for this hypothesis.

The ankylosaurs (fused-bone reptiles) were also long-bodied, small-headed herbivores, a bit shorter legged than the stegosaurs. Like the aetosaur thecodonts that lived millions of years earlier, the ankylosaurs evolved a covering of small bony plates that completely enclosed the body and presumably served as armor for protection against the giant theropod predators. Some species of ankylosaurs developed tails that resembled heavy bony clubs studded with short spikes. These would have been powerful defensive weapons.

The fourth group of ornithischians, the ceratopsians, or horned dinosaurs, were relatively large headed dinosaurs. Most had horns on the nose and above the orbits, and the posterior skull bones extended back to form a large "shield" over the neck (see fig. 12.4). They had relatively large, laterally compressed beaks and posterior teeth modified to form a unique shearing dental battery. The head shields are usually considered to have evolved as protection for the neck region, but they are relatively thin and in some species are perforated by large openings, which is not the expected optimal design for armor. More recent functional analyses suggest that they served to extend the area available for jaw muscle attachment or perhaps evolved for intraspecific display. Some modern chameleon lizards have frills with jaw muscles extending onto them, which they use as display structures. The dental battery provided a self-sharpening, continually growing, vertical shearing device, which optimal design analysis suggests is a specialization for slicing tough vegetation.

We've now looked at the major groups of dinosaurs, animals that because of their gigantic size and unusual morphologies have long captured the imaginations of both the public and scientists alike. Until recently, dinosaurs were thought to have been sluggish animals because they were assumed to be ectothermic. Their body temperature and metabolic rates were thought to be lower than endotherms and fluctuate with changes in local air temperature. This interpretation was based on extrapolation from modern reptiles, with little consideration given to the ways in which dinosaurs might have differed from their surviving relatives. In the early 1970s, a few paleontologists took a fresh look at the evidence of dinosaur activity levels and concluded that they were endo-thermic, active animals, similar in metabolic rate and activity levels to birds and mammals. This new interpretation sparked a decade of contro-versy. The main lines of evidence involved limb morphology, bone histology or microstructure, and assessments of feeding levels (preda-tor/prey ratios and herbivore ingestion rates).

To review, functional analysis of dinosaur limb bones indicates an erect rather than a sprawling posture, more like what we see in modern birds and mammals rather than in modern reptiles. From limb propor-tions and inferred stance, sauropods have been compared to elephants, ceratopsians to rhinoceroses, and the smaller bipedal carnivorous dino-saurs to flightless birds. From such comparisons, it has been inferred that dinosaurs had mammal-like (or birdlike) locomotor activity. It has then been argued that such a high locomotor activity level would have required a high metabolic rate and endothermy. However, recent physi-ological studies comparing energy use during locomotion in animals with sprawling versus erect posture found no difference in the energetic cost of locomotion.

The bone histology of dinosaurs is similar in some ways to that of mammals and birds and differs from that of most modern reptiles and amphibians. These differences are correlated with high growth rates. It has been argued that high growth rates require relatively constant high metabolic rates, and the similarity in bone type has been taken as evi-dence of endothermy in dinosaurs. High growth rates are related to in-creased metabolic rate. But increases in metabolic rate can occur without endothermy. The very large size of most dinosaurs would have resulted in passive thermoregulation. Particularly in the warm, equable climates of their time, the dinosaurs, by virtue of their high body vol-ume to surface area ratios, would have been slow to cool off (or warm up) and thus would have maintained a relatively constant body temperature.

Endotherms must consume food at a rate five to ten times greater than ectotherms to support their higher metabolic rates. The small heads of sauropod dinosaurs seem too small to have allowed them to ingest the sheer volume of plant materials they would probably have required to maintain a warm-blooded herbivore their size, and thus it has been con-cluded that the sauropods must have been ectothermic with low meta-bolic rates like those of modern reptiles. However, a comparable analysis has not yet been done for the ornithischian herbivorous dinosaurs. On the other hand, some estimates of the biomass ratios of carnivorous dinosaurs to their potential prey are very low, suggesting that each individual theropod needed a large number of prey individuals to maintain itself and hence must have been endothermic. However, such estimates of fossil community structure involve many untestable assumptions with regard to possible preservational biases, and the predator/prey ratio arguments have been challenged as unreliable.

One final argument concerning dinosaur activity levels has been based on relative brain size. In recent years, it has been determined that, while most dinosaurs had the brain size one would expect for reptiles of their body size, some of the small theropods had relatively larger brains, overlapping the low end of the bird/mammal brain size range. Relative brain size is loosely correlated with metabolic rate. From this it has been inferred that at least those relatively large brained, small carnivorous dinosaurs had metabolic rates and activity levels like those of modern birds and mammals and must therefore have been endothermic.

Much controversy originally centered on the question whether dinosaurs were endothermic or ectothermic. Additional work has shown that there are no good data to indicate that dinosaurs were endothermic. There is, however, good evidence from bone histology that at least some dinosaurs had high growth rates. And it seems that while some dinosaurs, at least the large sauropods, were probably like modern reptiles, others, particularly the small carnivorous saurischians, were more like modern birds and mammals. Still others may have had metabolic rates neither typically reptilian nor typically bird/mammalian but rather something intermediate between the modern tetrapod extremes.

This extraordinary diversity of reptiles dominated land faunas from about 200 million to 65 or 70 million years ago (December 17–25). However, there was much more to the great evolutionary radiations of the reptiles of those times because they also invaded the oceans and took to the air, and we will turn to those developments in later chapters. But first we will complete our survey and look at some additional, more minor land and amphibious radiations that are of special significance because they produced most of the surviving groups of reptiles.

Lesser Land and Amphibious Radiations

CHAPTER 13

Rhynchosaurs

About 215 million years ago (December 15), shortly before the beginning of the great evolutionary radiations of the dinosaurs, an unusual group of reptiles had a brief period of ecological success. These were the rhynchosaurs (beaked reptiles), small- to medium-sized, relatively heavily built reptiles, with the most unusual cranial modifications. The skulls (fig. 13.1) were relatively short, deep, wide, and massively constructed. The front of the snout was extended into a large toothless beak, with a tusklike upper portion overhanging the lower jaw. Behind the beak, large numbers of small teeth were fused into two broad, flat tooth plates above, each with a groove into which the almost toothless lower jaws fitted. There were two large temporal openings behind the orbit, and there was space between the braincase and outer skull bones for an enormous mass of jaw musculature. These skull modifications are roughly reminiscent of what we saw in the dicynodont therapsids (fig. 11.4) and appear to be adaptations for handling a tough herbivorous diet. Rhynchosaur limbs and girdles were modified only slightly from the primitive reptile condition. For a brief period of time rhynchosaurs were the most common forms found in many reptile land faunas, and then they disappeared from the face of the earth.

Turtles

Another group that appeared about the same time as the rhynchosaurs, but was more successful with respect to longevity, was the turtles (order Chelonia). Like the rhynchosaurs, turtles display significant cranial modifications, but their postcranial transformations have been truly drastic. Early in their evolutionary history, turtles lost all their teeth, developed a sharp beak covered by a horny sheath, and fused the bones

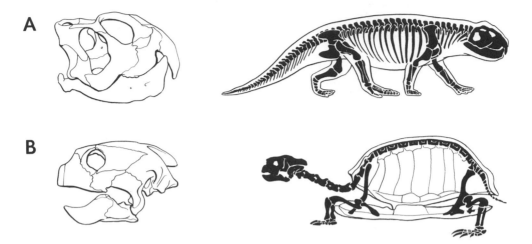

Fig. 13.1   *A*, skull and body outline of the rhynchocephalian, *Scaphonx*, about 3.5 feet long (after Romer, 1956). *B*, skull and shell of a modern turtle (after Gregory, 1951; Huene, 1956).

of their palate to make an extremely strong jaw apparatus. The sharp-edged horny sheath covering the beaked jaws makes an efficient piercing and shearing device and serves modern turtles with a wide variety of diets. Unlike other advanced reptiles, turtles never developed perforations in the outer skull wall behind the orbit, but they achieved the same functional result, ultimately greater space for large jaw muscles, by emargination of the outer skull wall from behind. In extreme cases, turtles have enlarged temporal fossae similar to those of advanced therapsids and mammals (fig. 13.1). Orbits are usually relatively large in turtles, and they also usually have an otic notch for a large eardrum.

While the skulls of turtles are unusually modified from the primitive reptile condition, it is the postcranial modifications that are extraordinary (fig. 13.1). The body is relatively short and enclosed in a rigid bony shell, covered externally by horny epidermal scales. The upper portion of the shell (the carapace) is composed of dermal bones, to the undersides of which are fused the neural arches of the shortened vertebral column and the ribs. The lower portion of the shell (the plastron) is formed in front by part of the pectoral girdle and posteriorly by what appear to be expanded belly ribs. The limb girdles are located inside the rib cage. The limb girdles are modified from the primitive tetrapod condition, locomotion and breathing are greatly constrained by the rigid shell, and the feet are relatively short and broad. The neck vertebrae are modified to allow withdrawal of the head into the shell, in a vertically oriented S curve in one group of living turtles, and in a sideways fold in the other group.

There are about 400 species of living turtles, which range in size from about 6 inches to a few feet in length. Some extinct turtles were giants, as long as 12 feet. Most turtles are freshwater aquatic species, but some are completely terrestrial (the land tortoises), and others are marine (sea turtles). Their rigid shells do not allow lateral movements of the body for aquatic locomotion, and in the stronger swimmers, such as the marine turtles, the digits are lengthened to support paddle-like hands and feet, which serve as the propulsive organs. Turtles eat a wide variety of diets, ranging from purely carnivorous (mainly fish and aquatic invertebrates), to omnivorous, to purely herbivorous (some land tortoises). There is a modest amount of variation in skull and shell construction in association with different diets and habitats.

Lizards

Lizards are a group of reptiles that have a long evolutionary history, but, with a few exceptions, they have not diverged significantly from the primitive reptile body plan. They first appeared around 225 million years ago (December 17) and apparently had a modest evolutionary radiation contemporaneous with that of the dinosaurs. There would have been little or no competition between lizards and dinosaurs, owing to the great size differences between them. The main cranial modifications in lizards were the development of two temporal openings in the skull, followed by the loss of the bony bridge across the bottom of the lower opening (fig. 13.2). The jaw joint is located at the back of the lower opening, and loss of that bar of bone allows for some movement of the upper bones of the jaw joint. Postcranially, the shoulder joint was modified to allow a greater range of movement of the humerus. Correlated with this change there is less rotation of the humerus about its long axis during the step cycle, and the humerus is more slender than in primitive reptiles.

Nevertheless, lizards maintain a sprawling posture during slow locomotion. Their pectoral girdle is modified from the primitive condition by the development of perforations where the main muscles attach, presumably analogous to the holes developed in the temporal region of the skull. The proposed functional significance of the holes is to allow for

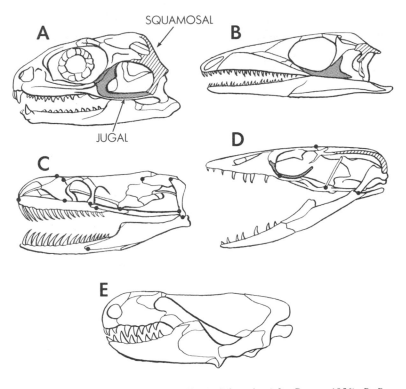

Fig. 13.2   Living and fossil reptile skulls. *A, Sphenodon* (after Romer, 1956). *B, Prolacerta*, an early lizard (after Colbert, 1980). *C, Python*, a snake (after Smith, 1943). *D, Varanus*, a modern lizard (after Frazetta, 1962). *E, Amphisbaena*, an amphisbaenid (after Romer, 1956). Dots show movable joints of the skull.

better attachment of muscles to bone. In some lizards, a movable joint develops between bones of the ventral plate of the pectoral girdle, allowing rotation of the girdle. This rotation adds significantly to front limb stride length as it increases the distance that the limb can reach.

There are about 3,000 living species of lizards, most of which are small (a few inches to a foot in length) and insectivorous. Some of the larger lizards are herbivores, and the very largest living lizard, the 8–10 foot Komodo dragon, is a carnivore that feeds on large mammals. Most lizards do not deviate much from the basic lizard body plan, but a few groups have evolved some interesting specializations. Old World chameleons can shoot their tongues out almost a body length in front of themselves to catch insects, and in addition they have pincer-like hands and feet (with opposable digits) and a prehensile (gripping) tail for slow, stalking locomotion on small branches. Some lizards developed very long, straight ribs that extended out from the body and supported a skin membrane up to a body length in width. This is a specialization for gliding among trees that appeared a few times early in the evolutionary history of lizards and can be seen in one living genus today. A common trend among lizards has been a reduction in the size of their limbs, often accompanied by an increase in body length. In extreme cases, limbs are totally lost and locomotion is accomplished by lateral undulation of the body. One group of ancient limbless lizards was enormously successful and produced a radiation almost as great as that of the lizards themselves. This group, the snakes, is discussed below.

Most lizards are (and were) small, with fragile bones, and therefore they are not particularly well preserved in the fossil record. One exception was the mosasaurs, a group of lizards that became secondarily aquatic, invaded the oceans, and, unconstrained by gravity, grew to gigantic sizes of 30–40 feet in length. Mosasaurs evolved paddle-like hands and feet, with elongate webbed digits, but their main propulsive organ was a long, laterally flattened tail (fig. 13.1).

Snakes

Snakes evolved from a group of limbless lizards about 100 million years ago (December 24) and underwent a very successful radiation of their own. Transformations from the standard lizard condition involved the total loss of limbs, a great increase in body length, increased complexity of intervertebral articulations and of vertebral column musculature, and a radical remodeling of the skull (fig. 13.2). Alterations in the skull included the loss of some bones and a freeing of others, so that virtually all the upper skull bones can be stretched apart from each other and the left and right sides of the jaw apparatus can move independently (fig. 13.2). These changes allow snakes to swallow prey larger than their own skull width, by moving first one side of the jaw and then the other forward over the prey, which is held in place by the snake's sharp, backwardly pointing teeth. Some snakes evolved specializations for killing prey by constriction and others by the injection of poison (an ability evolved several times among snakes). Locomotion is accomplished by lateral undulations of the body, enhanced by the large number of complex vertebrae and complex vertebral musculature.

We can view the cranial modifications that allow ingestion of relatively large prey as the key adaptation responsible for the emergence of snakes as an independent radiation from lizards. Because most lizards are relatively small and insectivorous, and most snakes are larger and carnivorous, there is little competition between the two groups. There are about 2,500 species of living snakes, ranging in size from about 8 inches long up to giants 35 feet in length. Most are terrestrial, but some are arboreal, burrowing, or aquatic, and among the aquatic snakes are both freshwater and marine species.

Amphisbaenians

Another small group of mostly limbless lizards are the amphisbaenians, which are burrowing specialists. Their skull is blunt or wedge shaped and drastically modified for pushing through soil (fig. 13.2). One of their unique features is a median tooth at the front of the upper jaw that fits between two lower teeth to form a nipping device. There are about 200 living species of amphisbaenians, most a foot or two in length, and they spend their lives underground, feeding on invertebrates.

Sphenodon

The last group of living reptiles is represented by a single species, found today only in New Zealand. Sphenodons (also known as tuataras), are lizard-like animals, 1–2 feet in length. These animals have two temporal openings in the skull and are less specialized than lizards in that they retain the lower bar in front of the jaw joint (fig. 13.2) and in lacking perforations in the pectoral girdle. Fossils are known from about 200 million years ago in North America and Europe that are very similar to the living *Sphenodon*.

Reptile Designs
for Marine Life

While some early archosaurs were evolving into the amphibious phytosaurs and crocodiles and others were giving rise to the early members of the great terrestrial dinosaur groups, several other groups of reptiles were becoming specialized for life in the oceans. Among the most spectacular and most specialized of these were the animals known as ichthyosaurs, plesiosaurs, and mosasaurs.

Ichthyosaurs

Ichthyosaurs (fish-reptiles) were the most fishlike of all the aquatic reptiles. They were sizable animals, ranging from 5 to 35 feet in length, with large elongate heads, very short necks, streamlined bodies, and a deep vertical tail fin like that of powerfully swimming bony and cartilaginous fishes (fig. 14.1). A downturned vertebral column extended into the ventral lobe of the tail (technically, a reversed heterocercal tail), but there is no evidence of fin rays or other internal supports for the equal-sized dorsal lobe of the tail fin. There was a large fleshy dorsal fin about midway along the body, and the fore and hindlimbs were transformed into short finlike flippers. These flippers evolved through extreme shortening of the limb bones and an increase in the number of phalanges (toe bones) and sometimes in the number of digits (fig. 14.1). Weight support is no problem in the water, and the vertebral column of ichthyosaurs was modified by loss of the interlocking zygapophyses that strengthen the vertebral column in land tetrapods. Ichthyosaurs had enormous eyes and long, narrow jaws that were lined with a uniform row of small, sharp teeth. The nostrils had moved back from the primitive position at the tip of the snout to come to lie on the top of the head, just before the eyes.

The fishlike body shape of ichthyosaurs suggests that they were powerful swimmers and also that they had become so specialized for

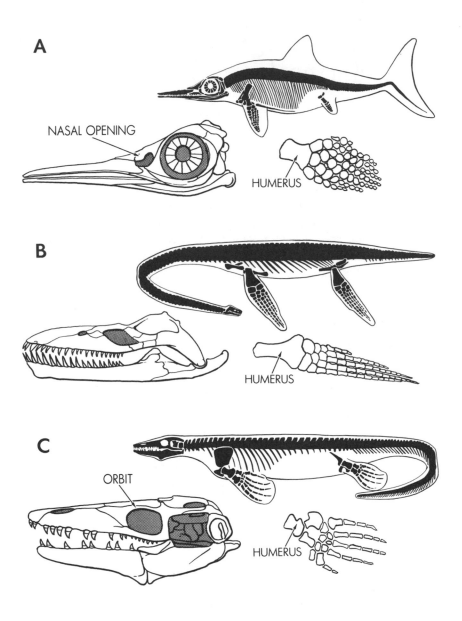

Fig. 14.1   Skull, body shape, and forelimb of marine reptiles. *A*, ichthyosaur (after Romer, 1966). *B*, plesiosaur (after Gregory, 1951). *C*, mosasaur (after Russell, 1967).

aquatic life that they could no longer come out of the water onto land. Since reptilian eggs must be laid on land, we would predict that ichthyosaurs gave birth to their young alive, a prediction confirmed by some rare fossil finds showing unborn fetuses within their mothers' bodies. The large tail fin was the main propulsive organ (via lateral undulations, as in fish), and the dorsal and paired fins provided stabilization and maneuverability in the water. Their large eyes suggest that vision was the ichthyosaurs' main sensory modality. Unlike the secondarily aquatic amphibians, aquatic reptiles never had a lateral line system. Their long, narrow jaws bearing a large number of similar-sized, long, conical teeth indicate that the ichthyosaurs were carnivorous. Parts of squid tentacles have been found in their stomachs.

Plesiosaurs

Plesiosaurs were large or very large animals, ranging in length from 10 to 50 feet, and their short, stocky bodies, short tails, and limbs converted to relatively long, large flippers gave them a unique shape (fig. 14.1). The lower portions of their pectoral and pelvic girdles were expanded, and the abdominal region between them was strengthened through the extensive development of a lower set of rodlike elements called gastralia, or belly ribs. This made for an extremely strong trunk, with large attachment areas on the undersides of the girdles for powerful muscles. There were two groups of plesiosaurs, one with small heads and very long necks and the other with large heads and relatively short necks. Both groups had relatively long jaws, lined with a uniform row of relatively narrow, conical teeth.

The two pairs of flippers were the main propulsive organs in plesiosaurs. Until recently it was believed that they paddled or rowed through the water, but a new functional analysis suggests that they also may have flown through the water by moving their flippers up and down, like penguins and sea turtles. This would account for the expanded ventral areas of the limb girdles, which indicate powerful limb depressor musculature for the propulsive downstroke. The long-necked, small-headed plesiosaurs probably swam among schools of small fish and used their long necks to swing their heads within reach of prey. The large-headed plesiosaurs appear to have been faster swimmers, and they probably pursued and caught larger prey.

Mosasaurs

About 100 million years ago, long after the first appearance of ichthyosaurs and plesiosaurs and near the end of the great Age of Reptiles,
a new kind of marine reptile appeared in the oceans. These were the
mosasaurs (fig. 14.1), large-headed, short-necked gigantic lizards (15–
40 feet long), which became secondarily specialized for marine life by
developing a long, laterally compressed tail and elongate digits that
supported webbed flippers. Lateral undulations of the tail were the main
source of propulsion, and the paired flippers were used mainly as stabilizers and for steering. The large jaws, lined with a uniform row of
sharp teeth, suggest that the mosasaurs fed on large fish and other soft-
bodied aquatic life.

Other Aquatic Reptiles

Several other groups of reptiles became secondarily aquatic, although
none of them showed the extremes of body modification seen in the
three groups we have just considered. The nothosaurs were small- to
medium-sized (2–5 feet long), slender reptiles with long necks and with
limbs only slightly modified in the direction of paddles. Their distal
limb bones were shortened, and their hands and feet were relatively
large. From their anatomy alone it is impossible to be sure that they
were aquatic, but the fact that their fossil remains are found only in
marine deposits is supporting evidence. From their small, narrow, conical teeth we can infer that nothosaurs fed on small fish. Form-function
correlations show that many modern fish eaters, such as the needlefish
in figure 8.4, have a similar dentition.

The placodonts were a very different group of aquatic reptiles.
Placodonts were medium-sized (5–10 feet in length), stout-bodied reptiles, whose outstanding specialization was the conversion of their dentition into a few large, flat crushing plates and the modification of their
jaw apparatus to produce powerful bites. There is no evidence that their
tails or limbs were modified for aquatic locomotion, and we infer that
they were aquatic because their fossil remains, like those of the
nothosaurs, are found only in marine deposits and their extremely modified dental apparatus is seen elsewhere only in shellfish eaters. Some of
the later species of placodonts developed a complete covering of dermal
bony plates, something like the shells of turtles.

The nothosaurs and placodonts lived only from about 225 to 200 million years ago (December 17), so we may view them as early experiments in aquatic life for reptiles, not nearly as successful, in terms of longevity and diversity, as the ichthyosaurs and plesiosaurs that succeeded them. Finally, in thinking about marine reptiles, we should remember that, in addition to the forms mentioned above, there were also the marine turtles and crocodiles mentioned in chapter 12.

Reptile Designs for Flight

The evolutionary modifications of the basic tetrapod body plan that we have seen in the great dinosaur radiations and in some of the marine reptiles were indeed extraordinary transformations. However, at least as remarkable are the evolutionary transformations for flight undergone by two different groups of small bipedal archosaurs. Requirements for flight are fairly severe for an animal, even one the size of a small vertebrate. They include a strong but flexible airfoil (flight surface or wing) with the potential to change shape; a light body; strong wing muscles; the ability to sustain a high level of muscle activity for relatively long periods, which in turn requires a relatively high and constant metabolic rate; and good muscular coordination. The transformations to meet these requirements occurred in the groups known as pterosaurs and birds, about 200 and 150 million years ago, respectively.

Pterosaurs

Pterosaurs (winged reptiles) were flying reptiles that ranged from the size of a sparrow to giants with wingspans as great as 35 feet! Their forelimbs were transformed into wings through the great elongation of the fourth digit, which supported the outer half of a tough flight membrane that was stiffened internally by elastic fibers. The clawed remnants of the first three digits projected forward from the leading edge of the wing, and medial to them was an additional projection, a unique bone called a pteroid, which apparently supported a small anterior flight membrane (fig. 15.1).

The shoulder girdle was modified to allow a powerful downstroke, and the sternum (breast plate) was enlarged to provide an attachment area for a large pectoralis muscle that provided the propulsive stroke. The body was relatively short and rigid, and the bones were thin walled and hollow to maximize lightness. The hindlimbs were similar to those

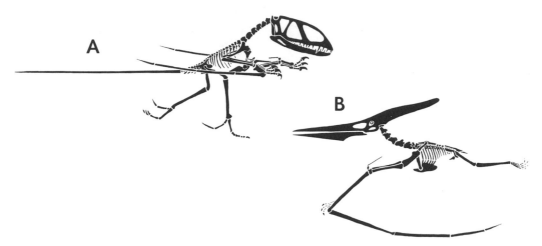

Fig. 15.1   Flying reptiles. *A, Dimorphodon,* with a wing span of about 4 feet (after Padian, 1983). *B, Pteranodon,* with a wing span of about 20 feet (after Eaton, 1910).

of small bipedal archosaurs, and recent functional analysis suggests that pterosaurs could walk bipedally quite well.

Casts of the inside of the braincase (called endocranial casts, or endocasts) reveal that pterosaurs had relatively large brains for reptiles, with a large optic tectum, small olfactory bulbs, and a large cerebellum. The size of the cerebellum suggests unusually good muscular coordination, and, among living vertebrates, large relative brain size is correlated with endothermy. Additional evidence that pterosaurs were endothermic comes from some rare fossil finds that preserve imprints of hairlike structures over the wings and body. These structures could well be the insulation expected in endotherms. Endothermy reflects a relatively constant high metabolic rate, one of the requirements for flight.

There were two main groups of pterosaurs, the earlier one composed mainly of small forms, with a long stiff tail and unreduced dentition. The later group, composed of larger forms, had lost the tail and often reduced the dentition (fig. 15.1). Pterosaurs had quite a diversity of tooth and jaw shape. Most of the small ones had small, sharp teeth and probably were insect eaters. The largest species were toothless and are thought to have soared far out to sea (their remains are often found far from ancient shorelines) to snatch fish from near the surface, much as many oceanic birds to today.

Because the wings of pterosaurs were composed of an uninterrupted sheet of membranous tissue, in contrast to the individually movable flight feathers of birds or the multiple braced (by digits) membranous wings of bats, the flight ability of pterosaurs was long considered to have been far inferior to that of bats or birds. The same logic led some scientists to suspect that pterosaurs also were incapable of powered flight, and to assume that they were gliders that had to climb up trees or cliffs and drop off to achieve takeoff velocities. However, more recent analyses of pterosaur skeletons, including the use of flying models in wind tunnels, suggest that they were indeed true fliers, capable of powered flapping flight and able to take off from the ground. Presumably they achieved fine control over the configuration of their wings by using the action of muscles on the stiffened fibers that extended throughout the flight membrane.

Paleontologists and biologists, in general, have tended to consider extinct groups inferior to surviving ones, and it has long been thought that birds outcompeted pterosaurs and were responsible for their eventual demise. However, there has been no analysis to support this belief, and the fact that pterosaurs and birds coexisted for almost 100 million years (the last of the pterosaurs died out about 70 million years ago) suggests that competition was not a major factor in the extinction of the pterosaurs.

Birds

Birds evolved from ancestors similar in form to those of pterosaurs, but they met the structural requirements for flight in somewhat different ways (fig. 15.2). Birds' flight surface was not a membrane attached to an elongate digit but rather was composed of a unique evolutionary feature—feathers—that attached to the arm bones and fused remnants of wrist and hand bones. Feathers are thought to have evolved from epidermal horny scales, but they are much more complex in structure. Flight feathers are composed of interlocking barbules that produce a strong, light, flexible surface, one that can be restored by preening after it is broken apart (fig. 15.3). Because they are individually movable, feathers allow for marvelously fine control over the configuration of the wings. The power stroke is forward and down, and asymmetrical flight feathers on the outer half of the wing provide forward thrust while feathers of the inner half of the wing provide lift.

The sternum of birds is greatly enlarged and has a prominent keel to provide sufficient attachment area for the large flight (pectoralis) mus-

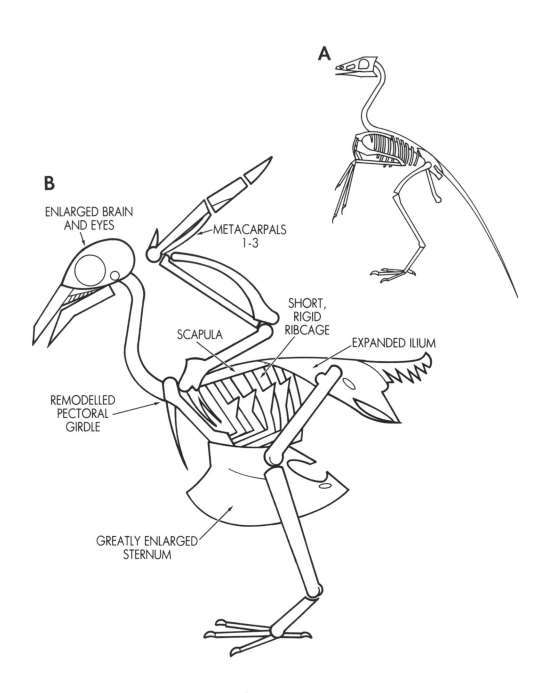

A

B

ENLARGED BRAIN
AND EYES

METACARPALS
1-3

SHORT,
RIGID
RIBCAGE

SCAPULA

EXPANDED ILIUM

REMODELLED
PECTORAL
GIRDLE

GREATLY ENLARGED
STERNUM

Fig. 15.2   Bird design for flight. *A, Archaeopteryx. B*, a pigeon.

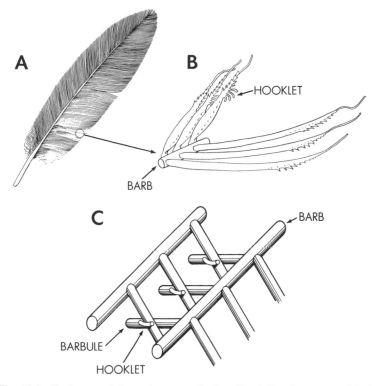

Fig. 15.3 Feather morphology. *A*, contour feather. *B* and *C*, enlargements of feather structure (after Hildebrand, 1982).

cles, which may constitute almost one-third of the total body weight of strong fliers. The pectoral girdle is greatly modified to provide strong bracing for the flight muscles, with the clavicles fused into a V-shaped "wishbone" (furculum) at the front end of the sternum. The rib cage is short and rigid, braced by overlapping processes between adjacent ribs.

The pelvic girdle is greatly expanded in birds, providing strong contact with the vertebral column and a large area for attachment of hind limb muscles. The hindlimbs are relatively large, similar in construction to those of advanced archosaurs and small carnivorous dinosaurs like *Deionychus* (chap. 12), and designed for efficient bipedal locomotion and use as "landing gear." The tail skeleton has been reduced to a stump, composed of the fused remnants of a few caudal vertebrae, and it provides a strong attachment area for large tail feathers (used in braking for landings). The bones of the skeleton are thin walled and hollow, designed for maximum strength at minimum weight, and the skeletons of large flying birds may weigh less than their feathers.

The skull of birds appears stripped to essentials: a relatively large globular braincase, large orbits, and slender, toothless jaws that are covered by a horny beak. The brain is five to ten times as large as that of amphibians or reptiles of similar body size, and it has a relatively large cerebellum (related to improved muscular coordination), large optic lobes, and small olfactory bulbs.

To power their flight apparatus, birds are endothermic, with a relatively high, constant metabolic rate six to ten times as high at rest as that of reptiles at the same body temperature. A covering of specialized down feathers provides extremely effective insulation and helps maintain the high body temperature. Birds have relatively large hearts, which are four chambered to provide for complete separation of oxygenated and unoxygenated blood. Their lungs have a complex system of air sacs that allow for a more efficient flow of air over the actual respiratory surfaces and also help dissipate heat (via expired air) generated by the heart and flight muscles. Most birds eat energy-rich food such as insects, seeds, fruit, and small vertebrates and have twice the blood sugar levels of mammals. Thus circulatory, respiratory, and digestive systems have been modified in birds to allow for the long periods of high activity rates required for flight.

These adaptations for flight are found in modern birds, but most were not present in the earliest birds. The oldest known fossil birds, from deposits 150 million years old, were robin-sized animals with skeletons extremely similar to those of small advanced archosaurs or tiny carnivorous (theropod) dinosaurs. In fact, the only skeletal feature that distinguishes *Archaeopteryx,* the first bird, from those reptiles is the presence of relatively longer digits on the hands. *Archaeopteryx* would have been classified as a reptile except for one feature: the clear imprints of feathers surrounding a few unusually well preserved fossil specimens. The absence of an expanded sternum in *Archaeopteryx* suggested to earlier paleontologists that it lacked the enlarged pectoralis muscles necessary for powered flight. They therefore concluded that *Archaeopteryx* was a glider. However, the feather impressions show asymmetrical vanes, such as we see only in flight feathers in modern birds, and the reconstructions of wing area fall within the range seen in modern flying birds of similar body size. Thus it seems likely that *Archaeopteryx,* despite its lack of skeletal adaptations for flight seen in modern birds, was indeed a true flier, capable of powered flight.

The switch from terrestrial locomotion to powered flight is a drastic change, and there has been much speculation about what the transitional stages were like. The traditional view was that birds went through an arboreal gliding stage before becoming true fliers. Gliding does not require wing flapping but depends on the ability of an animal to minimize its angle of descent. Animals that have a pathline from take-off to landing of less than 45° to the horizontal are called gliders.

A few years ago researchers pointed out that the enlarged hands seen in *Archaeopteryx* are not like those of modern arboreal gliders. Living gliders have membranes supported by elongate ribs (some lizards) or stretched between equal-sized fore- and hindlimbs (various mammals). An alternative hypothesis was suggested, in which the hands became elongated in the bipedal ancestors of birds to enhance their abilities to flush and trap insects from low vegetation. According to this hypothesis, feathers evolved initially for insulation or to increase the surface area of hands and arms available to trap insects. Such increased surface areas could function to increase the length of the floating phase in bounds taken by a small bipedal runner, and enhancement of floating phases may have provided the transitional stages between bipedal running and flight.

These two ideas can be called the "trees down" and "ground up" theories for the origin of powered flight. Both theories propose that the animals go through an intermediate gliding stage. Current controversy centers around whether gliding is a feasible intermediate stage in the evolution of powered flight. Are the morphological correlates of gliding the same features that could be important for powered flight? This question arises because the morphology of living gliders such as flying squirrels is so different from the morphology of the fossil *Archaeopteryx*. To examine this question we have to begin by reviewing how animals glide or fly.

Just as with swimming, an animal traveling through the air experiences drag. The animal is also pulled downward by gravity unless there is some upward force to counter its weight. That counterforce is lift, which acts at a right angle to drag. Lift in the air is generated just as it is in water: air traveling at different velocities above and below the wing or body creates a pressure differential (remember Bernoulli's principle). It is no coincidence that the cross section of an airplane wing has the same streamlined shape as the profile of a fish. When that shape, called an airfoil, is positioned at a slight angle to the airstream, it maximizes the lift-to-drag ratio. Maximizing that ratio is how gliders minimize their angle of descent and maximize the horizontal distance they cover in their descent.

Mathematical analysis shows quite nicely that morphological features such as an enlarged hand with feathers are important changes that improve gliding distance by increasing the lift-to-drag ratio. Morphological features for gliding and powered flight are not mutually ex-

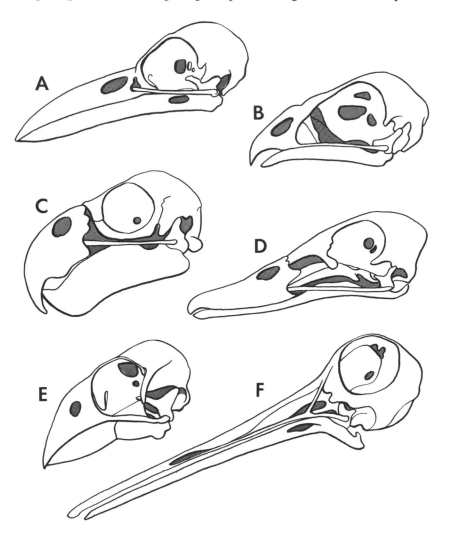

Fig. 15.4   Bird skull diversity. *A, Corvus* (crow). *B, Circus* (hawk). *C, Amazona* (parrot). *D, Anas* (duck). *E, Eophona* (hawfinch). *F, Scolopax* (woodcock).

clusive. The analysis does not settle the question whether active flight evolved from the "trees down" or from the "ground up," but it does confirm the likelihood that gliding was an intermediate step in either case.

The fossil record of birds is sparse, owing to the very fragile nature of bird bones. Nevertheless, enough is preserved to indicate the presence of species with essentially modern anatomy by about 20 million years after the appearance of *Archaeopteryx,* and the existence of several of the modern orders of birds a relatively short time later. Most fossil birds can be assigned to modern orders, and we have evidence of only one type of extinct bird that was very different from any bird living today. These were giant flightless birds, standing 5–7 feet high, with large clawed hind feet and very large tearing beaks. They lived during the early part of the Age of Mammals (the last 65 million years) in North America and for most of that period in South America, and they appear to have been among the top terrestrial predators in their respective faunas.

There are about 8,700 species of living birds, and more than half of them belong to the order Passeriformes (perching birds, mostly song birds). There are twenty-seven other orders of modern birds, representing a great diversity of body types, from tiny hummingbirds to giant flightless ostriches, and they inhabit a great diversity of environments. Beak shape and foot structure have been greatly transformed in many groups of birds, and these features are specializations for various types of diet and habitat (fig. 15.4).

The Origin of Mammals

Mammals first appeared around 190 million years ago (December 16), descended from a line of small cynodont therapsid reptiles. They differ in a host of ways from their reptilian ancestors, and almost all of the differences appear to be reflections of a more active life, supported by a relatively high, constant metabolic rate.

The structural modifications involved in the origin of mammals reflect improvements in locomotor and feeding systems, and various stages in the transformation can be traced through 25 million years of therapsid reptile evolution. Mammalian locomotion involved a change in limb posture from sprawling to semi-erect, with elbows pointing out and backward, and knees pointing out and forward (fig. 16.1). As discussed in chapter 12, this change placed the hands and feet closer to the body's center of gravity, resulting in less need for large ventral limb muscles just to hold the body up off the ground. The limb girdles reflect this changed biomechanical situation. The only remnant in the pectoral girdle of what was once a large ventral plate is the rod-shaped clavicle, or collar bone. The upper portion of the pectoral girdle, the scapula or shoulder blade, was enlarged anteriorly to provide attachment area for additional musculature that strengthens the shoulder joint and helps rotate the limb antero-posteriorly. In the pelvic girdle, the upper portion (ilium) rotated forward, and the lower portion (pubis and ischium) rotated backward, so that the long axis of the pelvis came to be more anteriorly oriented. This produced changes in muscle direction to a more fore and aft orientation, resulting in enhanced mechanical efficiency in rotating the limbs forward and backward. (Dinosaurs achieved a similar result by expanding the ilium and modifying the shape and position of the pubis and ischium.)

The long bones of mammalian limbs became more slender in conjunction with the shift to a more erect posture. Bony processes developed at elbow, hip, and ankle joints to lengthen in-lever arms (see fig.

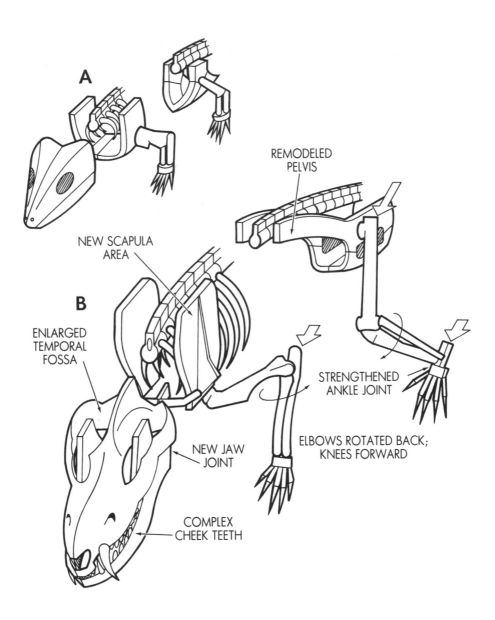

A

REMODELED
PELVIS

NEW SCAPULA
AREA

B

ENLARGED
TEMPORAL
FOSSA

STRENGTHENED
ANKLE JOINT

NEW JAW
JOINT

ELBOWS ROTATED BACK;
KNEES FORWARD

COMPLEX
CHEEK TEETH

Fig. 16.1 Basic mammal design. *A, Eocaptorhinus*, representing a primitive reptile morphology. *B, Didelphis* (the opossum), a mammal.

16.1) and improve the mechanical efficiency of the muscles that rotate the limbs backward at those joints. At the elbow, this elongation is called the olecranon process, or funny bone of the ulna, and it is the site of attachment of the triceps muscles. The femur developed a greater trochanter above the hip joint for attachment of the gluteal muscles from the ilium; at the ankle, the tuber calcis developed from the heel bone (calcaneum) for attachment of the gastrocnemius muscle, via the Achilles tendon. Finally, the ankle joint was remodeled so that a process from the heel bone (calcaneum) lay under and helped to brace the ankle bone (astragalus), making for a stronger ankle joint.

Ribs were lost from the region of the vertebral column between the head and shoulders (cervical vertebrae, numbering seven in almost all mammals) and for a short distance anterior to the pelvis (lumbar vertebrae). This change may have allowed for increased vertebral column flexibility. In addition, the first two cervical vertebrae, called the atlas and axis, were remodeled to provide for both strength and flexibility at the articulation of skull with the vertebral column.

Major changes in teeth, jaws, and jaw musculature modified the feeding system. The mammalian dentition differentiated into small incisors, large canines, and increasingly complex premolars and molars; the latter two types constitute the cheek teeth. Increased complexity of cheek teeth involved the growth of additional cusps on the crowns, with precise interlocking of cusps and ridges between upper and lower teeth. The more elaborate cheek teeth crowns provided mammals with shearing and crushing capability during chewing rather than the simple puncturing of single-cusped reptile teeth. As crown complexity increased, the cheek teeth developed multiple roots for firm anchorage. Teeth were no longer continuously replaced throughout life; instead, only two sets of teeth ("milk teeth" and adult teeth) appeared. This shift from continual tooth replacement to two sets of dentition was probably related, in part, to the evolution of precise occlusion. With continual tooth replacement, it is difficult to maintain close alignment between upper and lower teeth over the entire length of the tooth row simultaneously. But, with only one period of replacement, sets of lower and upper teeth will be in precise occlusion over virtually the entire life of the animal. Additionally, mammals only grow for a limited period of their lives, while most reptiles continue growing throughout their entire lives. The shift from continual tooth replacement by mammals may be related to the evolution of this general pattern of limited growth.

The changes in jaw structure involved expansion of the anterior-most and largest bone of the reptile jaw, the dentary, and the reduction in size of the more posterior jaw bones. In mammals, the jaw is composed of just the dentary, which has developed a condyle and made a new articulation with the skull at the squamosal bone (fig. 16.2). Thus mammals are said to have a dentary-squamosal jaw joint. The bones of the old reptile jaw joint, the quadrate and articular, have been reduced to tiny vestiges in mammals, and have come to lie in the middle ear cavity as incus and malleus, interposed between eardrum and stapes. This may seem like a far-fetched transformation, from reptile jaw joint to mammal ear ossicles, but the middle ear cavity lies just behind the jaw joint, so there was not much distance for the bones to travel. Furthermore, the transformation can be seen in the embryological development of modern mammals (fig. 16.2), and intermediate stages are preserved in some adult fossil therapsids, including one genus *Diarthrognathus* (two jaw joints), in which both the old reptile and new mammal jaw joints function simultaneously!

The functional significance of these remarkable changes was two-

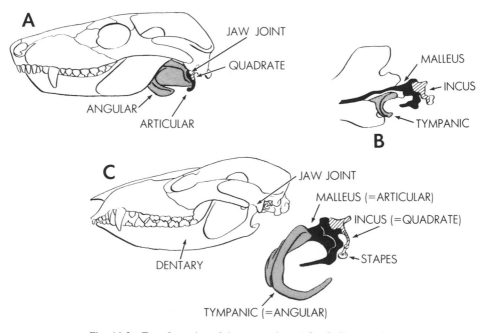

Fig. 16.2 Transformation of the mammal ear (after J. Hopson; Crompton and Parker, 1978). *A*, the skull and middle ear of a mammal-like reptile, *Thrinaxodon. B*, the lower jaw of an opossum fetus. *C*, the skull and middle ear of an adult opossum.

fold. First, the jaw was strengthened, because it was now composed of a single bone rather than several. Second, with three ear ossicles rather than one, mammals had the benefit of better amplification of sound vibrations from the tympanic membrane, since there were two additional joints and levers in the sound-transmitting system between eardrum and inner ear.

A horizontal flange of bone grew inward from each side of the upper jaw. These flanges eventually met in the midline and formed a secondary palate, separating the air passageway above from the mouth cavity below. This secondary palate strengthened the upper jaws by acting as a brace between its left and right sides and allowed breathing and chewing at the same time. As you may recall, a similar secondary palate evolved independently in crocodiles (chap. 12) and therapsids (chap. 11).

The changes in jaw musculature involved an increase in the overall size of the muscles and their increased differentiation (fig. 16.3). As the upper portion of the jaw musculature (the temporalis) grew larger, the

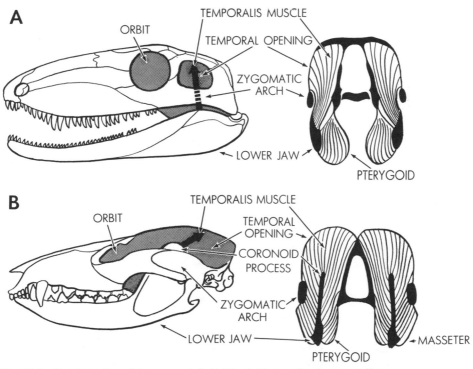

Fig. 16.3 Transformation of the mammal skull (after J. Hopson; Barghusen and Hopson, 1979). *A*, lateral and cross-sectional views of the skull of a primitive mammal-like reptile. *B*, lateral and cross-sectional views of the skull of a mammal.

single opening in the skull roof behind the orbit also enlarged, until, with the temporalis muscle expanded to a relatively enormous mass, all that was left of the old outer wall of bone external to the braincase was a strip along the lower border, the zygomatic arch or cheekbone (fig. 16.3). The temporalis muscle attached to the lower jaw via a vertical flange of bone, the coronoid process, which extended upward just in front of the jaw joint. This configuration changed the average direction of pull of the temporalis from straight upward to upward and backward (fig. 16.3, *A* and *B*). A lower and outer portion of the temporalis expanded downward to form a new major jaw muscle, the masseter, that extended from the zygomatic arch to the outer side of the back of the lower jaw (fig. 16.3). This new, expanded region of the lower jaw where the masseter attaches is called the angular process and is unique to mammals.

The old reptilian pterygoid muscle formed a third major portion of the mammalian jaw musculature. Called the medial or internal pterygoid in mammals, this third portion of musculature extended from the lower part of the front of the braincase to the inner side of the back of the lower jaw. The masseter and medial pterygoid form a sling that allows medial movement of the lower jaw and rotation about its long axis. The temporalis and outer portion of the masseter form a biomechanical couple. When they contract, they produce parallel forces on the lower jaw in opposite directions. This causes the jaw to rotate about its condyle with minimal loading (reaction force) at the jaw joint. Together these changes in jaw musculature and jaw configuration allowed for stronger chewing, more extended chewing without damage to jaws or joint, and more precise occlusion of upper and lower teeth.

These changes suggest that the earliest mammals were more agile and active than their primitive therapsid ancestors, with a more efficient digestive system (thorough chewing of food results in smaller particles to be swallowed, which have increased surface area and thus are more readily digested). Those differences are related to the relatively constant, high rate of metabolism that distinguishes modern mammals from modern reptiles and that results in mammals being endothermic. Maintenance of a high rate of metabolism allows sustained high rates of activity and requires a high rate of food intake and an efficient digestive system. In addition, there were changes in mammalian circulatory, respiratory, and excretory systems. As was the case with birds, the mammalian heart developed four chambers, creating a complete separation of oxygenated from unoxygenated blood. The surface area of the lungs expanded as the main chambers subdivided, and a sheet of muscle, the diaphragm, developed across the back of the rib cage and supplemented the rib cage muscles in the work of expanding and contracting the chest cavity. The excretory system improved through the development of a much more efficient kidney that was able to conserve water by producing a very concentrated urine. The mammalian kidney is also more efficient than the reptilian kidney in waste removal and the maintenance of proper blood chemistry.

The maintenance of a high metabolic rate requires good insulation, particularly in a small animal with a high ratio of surface area to body mass. Natural selection replaced the old reptilian epidermal scales with a mammalian covering of hair, which is an excellent insulating material. Like feathers, most hairs are equipped with muscles that allow erection and flattening, and these changes in position alter the amount of air trapped close to the body and the degree of insulation the hair affords. To aid in cooling the body, some mammals developed sweat glands in the skin. Early in mammal evolution some hairs developed special tactile functions. Called vibrissae, these large, stiff hairs are usually located in clusters on the face and are equipped with extensive sensory receptors at their base.

Mammals are generally thought of as live bearers, but judging from the condition in the two most primitive living mammals, the echidna and platypus, the earliest mammals probably were egg layers. It was not until some time later, but at least by 100 million years ago (when marsupial and placental mammals diverged from each other), that the reproductive system was modified to allow retention and nourishment of growing embryos within the mother's body. Many fish, a few amphibians, and some reptiles give live birth, but in some of those cases animals accomplish it by retaining eggs that then hatch inside the mother's body. Live-bearing placental mammals developed the ability to nourish the embryo from the maternal blood stream via a structure called the placenta. After birth, the young of all mammals, including the egg-laying echidna and platypus, are nourished by special secretions, called milk, produced by the female mammary glands. One of the main differences in reproductive systems between marsupials and placentals is that marsupial young are not nourished by a true placenta. The young of marsupials are born at a relatively early stage and finish development attached to a nipple in a pouch on the mother's abdomen.

The other major transformation in the evolution of mammals from reptiles was a five- to tenfold increase in brain size relative to body size. As was the case in birds, the cerebellum was relatively enlarged, perhaps reflecting a higher degree of motor coordination, but, unlike birds, early mammals had a fairly small visual system and the olfactory bulbs were relatively large. Further, mammals expanded a portion of the outer layer of the cerebral hemispheres of the forebrain into a major higher-information-coordinating center, called the neocortex or "gray matter." Many groups of later mammals experienced a further increase in relative brain size that was correlated with the increased size of the neocortex, compared with the early mammal condition. Most workers consider increased brain size an indication of increased "intelligence." However, while such a conclusion may seem intuitively correct, there are virtually no scientific data to support it, and the functional significance of increased relative brain size remains an unsolved puzzle. The puzzle is particularly intriguing because increased brain size occurred dozens of times independently among various groups of birds and mammals.

The changes in the mammalian skeleton can be seen in various stages in different lines of therapsid reptiles during the latter part of their history, and these changes are a grand example of parallel evolution. Only one line achieved all of the changes and gave rise to mammals. The earliest mammals were very small animals, mouse to rat sized. Their large olfactory bulbs, prominent facial vibrissae, improved sense of hearing, and relatively small eyes suggest that these mammals were nocturnal. This interpretation is based on extrapolation from living mammals with the same balance of sensory facilities. Their small size and the shape of their teeth, compared with the body size and tooth shape of modern mammals, imply that the earliest mammals were insectivorous.

Shortly after their first appearance around 190 million years ago (about December 16), mammals differentiated into four small groups, three insectivorous and one that may have been partly herbivorous, judging from its flattened, multiple-cusped teeth. This last group, known as multituberculates (from the many cusps on their cheek teeth), survived until about 35 million years ago. Of the remaining insectivorous groups, one line, which left no intervening fossil record, gave rise to the modern egg-laying echidna and platypus, which are known as the monotremes. Another line gave rise to all the other modern mammals and about 100 million years ago divided into the two main living groups, marsupials and placentals. For about 115 million years, mammals were small, insignificant elements in land faunas, completely overshadowed in terms of size, numbers, and diversity by the great reptile radiations. Then, from 65 to 70 million years ago, during the last of the Cretaceous Period, there was a wave of extinctions that decimated most of the reptile groups, as well as many kinds of invertebrates. After this wave of extinctions, the mammals underwent a rapid and massive evolutionary radiation, apparently filling the ecological niches that were vacated by the extinction of so many groups of reptiles. This initial wave of radiations, about 65 million years ago, produced many archaic groups of mammals, most of which are now extinct. About 10 million years later, another series of radiations produced most of the twenty living orders of mammals.

Today most of us think of endothermic mammals as superior to ectothermic reptiles, and therefore we may find it difficult to understand how, for so many millions of years during the Age of Reptiles, mammals were apparently outcompeted by reptiles and confined to a small number of nocturnal ecological niches. Two considerations may offer explanations. First of all, being endothermic is advantageous only in relatively cool climates or during night-time cool temperatures. For most of the great Age of Reptiles (225 to 65 million years ago), climates were warmer and less seasonal (cold/warm) than they are today, which meant that ectothermic reptiles would have had warmer body temperatures. Second, it is costly to be endothermic and maintain a high constant metabolic rate. Mammals require ten to thirteen times as much food energy as reptiles or amphibians to maintain the same body mass. This means that much less of their energy budget is available for growth and reproduction. Thus it may be more realistic to view ectothermy and endothermy as alternative strategies for dealing with life, each with its own advantages and disadvantages, depending on the environment. For the past 65 million years, the advantages of endothermy seem to have been dominant, for during this time, the so-called Age of Mammals, mammals and birds have come to dominate the land faunas. We will turn now to review the great diversity of body forms evolved by mammals.

Diversity of Mammalian Design

chapter 17

Skulls

It can be useful to look at skull diversity in terms of functional components, groupings of bones that serve a common function. On the most general level, skulls can be divided into two major parts: cerebral skull and facial skull. The former, sometimes known as the neurocranium, includes the braincase and sensory capsules (orbits and auditory bullae). The latter, also known as the splanchnocranium, includes the jaws and associated structures, like bony scaffolding for the attachment of jaw muscles (see fig. 17.1). The different functional components of the skull can evolve separately from each other, in response to different selective pressures, and the differential enlargement or reduction in size of various functional components can result in skulls of very different appearance.

We can compare the skull proportions of an opossum, representing the primitive mammalian condition, with those of other mammals selected to show unusual enlargement or reduction of various functional components (fig. 17.1). The horse, like most large ungulates, has a relatively large facial skull component (fig. 17.1), which houses a large dental battery for grinding tough vegetation. The anteater (fig. 17.1) has completely lost its teeth and reduced its jaw muscles, so its jaws are slender and reduced. Many lines of mammals evolved enlarged brains, and the extreme case, seen in modern humans, has resulted in a skull greatly distorted by its enormous braincase (fig. 17.1). The most extreme example of large eyes among mammals is seen in the tarsier, a small nocturnal primate, with orbits so large the jaws have been deflected downward to make room for them (fig. 17.1). Finally, the best examples of enlarged auditory bullae can be seen in many groups of desert rodents, which are nocturnal to avoid the heat of the day and have elaborated their auditory senses to detect predators at night. The

151

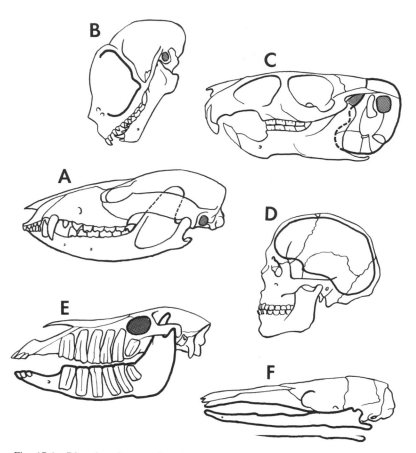

Fig. 17.1   Diversity of mammalian skulls, showing enlargement or reduction of functional components. *A*, opossum. *B*, tarsier. *C*, chinchilla. *D*, human. *E*, horse. *F*, anteater.

chinchilla (fig. 17.1) provides a modest example of this bulla enlargement.

It is commonly believed that the length of the snout of mammal skulls reflects nasal capsule size and olfactory acuity, but this is not true. The olfactory receptors in mammals are laid out on a folded, scroll-work arrangement of paper-thin bones called turbinals. They are located within the snout, but most of the area of the turbinals is devoted to air conditioning—cooling or warming and adding humidity to inhaled air and filtering out dust. The surface area of turbinals is increased by tighter folding of the turbinals, not by lengthening of the snout.

Snout proportions are controlled primarily by the requirements of the jaw apparatus for housing teeth, transmitting biting stresses, and so on. A more accurate indicator of olfactory acuity is the size of the olfactory bulbs, but they rarely have an impact on skull proportions.

In addition to differences in skull proportions from selection for differential enlargement of particular functional components, we also see differences resulting from evolutionary changes in body size. Cerebral skull and facial skull components scale differently with changes in body size. Facial components keep pace with evolutionary increases in body size, but cerebral components do not. As a result of this difference in scaling, the braincase, orbits, and bullae are relatively smaller in skulls of large species compared with their smaller relatives (fig. 17.2). This phenomenon of changes in shape with changes in size is known as allometry, and it accounts for much of the differences in skull shape that we see across mammals of different sizes.

Jaws

We see two major jaw shapes among mammals. In the primitive condition, seen in insectivores and carnivores, the jaw joint or articulation is at the level of the tooth row, with a large coronoid process (to accommodate a large temporalis muscle) and a small angular process (to accommodate small masseter and medial pterygoid muscles). In the derived condition, which occurs in ungulates and other medium-sized to large herbivores that grind their food, the condyle is raised above the

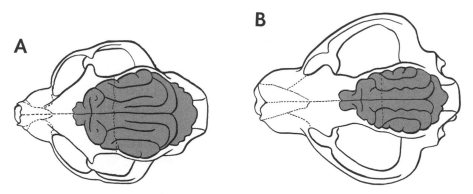

Fig. 17.2   Effects of scaling on skull shape; brain is colored black. *A*, cat. *B*, tiger.

level of the tooth row, and there is a small coronoid process and a large angular process, corresponding to reduction of the temporalis and expansion of the masseter and medial pterygoid muscles (fig. 17.3). The configuration that we see in ungulates evolved at least fifteen times independently in mammalian history and apparently represents a common solution to a biomechanical problem—the need for fine control of jaw movements in prolonged grinding of tough vegetation. The masseter and medial pterygoid form a sling on the same side of the jaw joint (the fulcrum of the jaw apparatus) as the tooth row (the position of the out-force). These muscles are therefore better situated than the temporalis for controlling jaw orientation and movement during the power stroke. There have been other modifications of the primitive mammalian jaw shape in various groups, but none as common as the ungulate type. We will discuss other unusual jaw shapes in chapter 18, when we mention the particular groups that possess them.

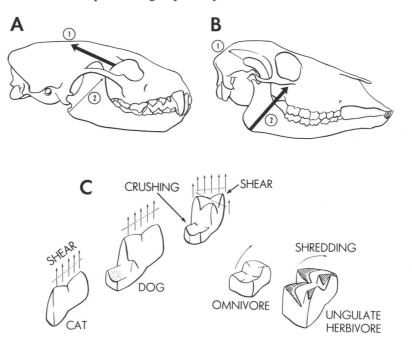

Fig. 17.3   Diversity in mammal skulls and teeth. *A,* carnivore skull. *B,* ungulate herbivore skull. Numbers indicate relative size of (1) temporalis and (2) masseter muscles. *C,* tooth types among mammals. Middle tooth represents the primitive condition. Arrows indicate direction of jaw movement.

Teeth

Teeth represent probably the greatest evolutionary diversification of the mammalian skeleton. Almost all major groups of mammals, and many minor ones, developed unique molar crown shapes, which can be recognized from a single tooth. This is indeed a boon for mammalian paleontologists, since isolated teeth are the most commonly found remains of fossil mammals. Because of the unique shapes of these teeth, it is often possible to identify even the genus and species from which they come. The evolutionary transformations of mammalian molar shapes affect the shearing, crushing, or grinding functions of the teeth and appear to be a response to the demands of different diets. However, teeth demonstrate quite clearly that there have been multiple solutions to common functional demands. Molar tooth patterns are often quite different among mammals with similar herbivorous diets.

Teeth can also provide clues to the jaw movements of an animal during feeding. As teeth move across each other, they produce scratch marks called striations on the tooth surface in the direction of the power stroke. Both fossil and modern teeth have wear lines that are visible under the microscope. The orientation of the striations reveals whether the power stroke of the lower jaw has a predominantly anterior, medial, or up and down direction of movement. In this way, we can correlate changes in jaw morphology in fossils with shifts in patterns of jaw movements.

We can see the primitive mammalian molar pattern in the opossum and modern insectivores such as hedgehogs and shrews. It consists of reversed interlocking triangles of high acute cusps, with a small low heel on the back of the lower molar (fig. 17.3). As the cusps and valleys of the upper and lower molars interlock, they perform puncturing, shearing, and crushing functions. With the evolution of carnivores specialized to eat larger animal prey than their insectivore ancestors, the shearing function of the tooth was enhanced by the enlargement and anteroposterior alignment of one set of shearing crests and usually the enlargement of a single pair of upper and lower teeth (called carnassials) that bear this crest. These modifications allow efficient, scissor-like shear along the long crests, in contrast to the cookie-cutter-like shear along short, oblique crests that we see in the primitive condition. Most carnivores retain the heel on the lower carnassial and some degree of crushing function, but in extreme cases, as in the cats, the heel is lost and the carnassial teeth function only for shearing.

In those mammals that shifted to omnivorous or herbivorous diets, such as pigs, bears, and humans, the molar cusps became low and blunt. The molar crown areas were enlarged and squared off by the addition of a fourth cusp on the upper molars and an increase in size of the heel on the lower molars (fig. 17.3). These changes resulted in a reduction of shearing function and an increase in crushing and grinding ability. In herbivores with diets of tough vegetation, such as leaves and grass, the four main cusps of the upper and lower molars usually became higher and developed complex anteroposteriorly elongate shapes. This produced steep ridges of enamel aligned perpendicular to the direction of jaw movement. Thus, when the enlarged masseter and medial pterygoids swung the jaw medially at the beginning of chewing, intricately folded ridges of enamel would grind past each other, forming a very effective shredding and grinding mill. In addition, many groups of herbivores (rodents, horses, deer) evolved high-crowned (and sometimes ever-growing) cheek teeth to compensate for the greater tooth wear caused by increasingly abrasive diets.

Modern elephants have uniquely modified cheek teeth to withstand the wear of an abrasive diet. Each cheek tooth is greatly enlarged through the addition of high, transversely oriented plates of enamel, with up to fifteen to twenty plates per tooth. During the life of the elephant, the teeth erupt and move into place sequentially and slowly, so that only one tooth is functioning at a time in each side of the jaw. When this tooth wears down, the next tooth erupts from behind. With transversely oriented enamel ridges on the cheek teeth, we would predict an anteriorly or posteriorly oriented chewing stroke, and indeed this is the case. Elephants have uniquely modified jaw musculature reflecting an anteriorly oriented power stroke.

Some groups of mammals specialized the anterior teeth. The most successful of these, the rodents, developed one pair of enlarged evergrowing incisor teeth. The enamel is confined to a strip on the front of the tips of the teeth. As a result of this configuration, when the upper and lower incisors occlude, they form a self-sharpening mechanism: the softer dentine at the tip wears down faster than the harder strip of enamel in front. With the development of gnawing incisors, rodents also modified their jaws and jaw musculature so that there were two working positions for the lower jaw: an anterior position for the incisors and a posterior position for the cheek teeth. These jaw positions are possible because, in rodents, the condyles are narrow and rest in longitudinal grooves. The lower jaws can then slide back and forth along the grooves.

While in most cases specialization for different diets involved increased complexity of cheek teeth, in some cases teeth became simplified. The cheek teeth of aquatic mammals that live on fish, like seals and porpoises, have become simplified through the reduction and loss of cusps. Many seals still retain vestiges of multiple cusps on their cheek teeth, but porpoises have single-cusped teeth, which are useful only for piercing and holding and which superficially resemble the teeth of their reptilian ancestors. And, as mentioned earlier, some mammals have lost their teeth altogether. At least four different lines of termite and anteaters—the pangolins, aardvarks, echidnas, and anteaters—have lost all vestiges of their teeth.

Limbs

In the primitive condition, mammalian limbs were relatively short, as in the opossum, with the distal segments (metapodials, tibia and fibula, and radius and ulna) short relative to the proximal segments (humerus and femur). Many lines of mammals became specialized for running, and in these groups the distal limb segments are elongated relative to the proximal segments. The biomechanical explanation for this shift is discussed in chapter 12. We can see extreme cases of this modification in some of the large ungulates such as deer and antelope (fig. 17.4). The limbs become oriented in a vertical plane, parallel to the body axis, and the joints, particularly those of the elbow and ankle, become hingelike so that the limbs can rotate only in a fore-and-aft direction. Elongation of the distal limb segments also brings a change in posture, from flat footed (plantigrade) to walking on tiptoe (unguligrade). Because the lateral digits are shorter than the middle ones, long-footed unguligrade mammals tend to reduce and lose the side toes, and the extreme case is seen in the modern one-toed horses.

As the lateral metapodials are reduced, the medial metapodials become the major load-bearing elements of the distal limb. In deer and their relatives (artiodactyls), the third and fourth metapodials are fused into a single element called the cannon bone. In horses, as mentioned above, there is only one remaining metapodial (the third), and it is

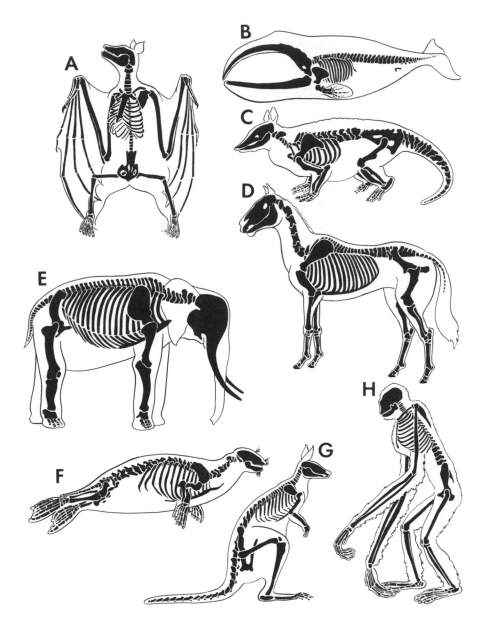

Fig. 17.4   Diversity in mammalian body shapes. *A*, bat, 4 inches long (after Young, 1962). *B*, whale, 40 feet long (after Young, 1962). *C*, armadillo, 2 feet long (after Gregory, 1951). *D*, horse, 8 feet long (after Young, 1962). *E*, elephant, 20 feet long (after Young, 1962). *F*, seal, 5 feet long (after Young, 1962). *G*, kangaroo, 40 inches long (after Gregory, 1951). *H*. gibbon, 20 inches long (after Young, 1962).

greatly enlarged. The single or fused metapodial is actually more efficient biomechanically than a parallel series of bones would be for providing strength against the large forces generated by the foot hitting the ground during running. Cross-sectional area is a critical parameter in determining the strength of a bone. As it turns out, for two bones to provide the same strength as a single bone, the two bones must have more combined cross-sectional area or be heavier than a single bone. Remember, the heavier the weight of the distal limb, the higher the moment of inertia and the more energy it takes to move the limb. A single distal limb bone is biomechanically more efficient than two bones because it can provide the same strength at a lower weight.

In addition, the foot of some runners shows another modification that is energetically efficient. In horses and artiodactyls, there are large ligaments that run from the metapodials to the phalanges. When the foot of the animal strikes the ground, the force bends the foot at the joint between these two bones. This stretches the ligaments that run across the joint. The ligaments are elastic (somewhat like rubber bands), and as they are passively stretched they absorb and store energy. As the foot subsequently is lifted off the ground, in the next step, that energy is recovered and applied to the next stride. No muscle contraction is directly involved in this action. Thus, not only is energy saved and reused by this spring mechanism, but there is no additional energetic cost involved in this form of energy storage.

No mammals come close in size to the larger dinosaurs. An elephant weighs only 7 tons while some sauropods were over 80 tons. Nonetheless, some of the larger living mammals are heavy enough that their limbs are modified for supporting heavy loads. Just as in the large dinosaurs (see chap. 12), the limbs of graviportal ("heavy to carry") mammals have shifted to under the body. When the limbs are aligned under the body, the weight of the animal exerts a vertical (compressive) force parallel to the long axis of the bone. If the limbs are not aligned vertically, the body exerts a bending force acting at a right angle to the bone, similar to the way that wind (a force) bends a large tree. Bones can resist vertical forces better than bending or perpendicular forces. By vertically aligning the limbs under the body, the bending forces are minimized. Also, as explained in chapter 12, muscle action could not support a body the size of these large animals. The vertical column of bone shifts the support role from muscle to bone.

Graviportal mammals are also characterized by longer proximal limb bones and shorter distal bones, the opposite situation to that seen in runners. Again, the short, stocky distal segments are more resistant to bending loads. The feet of graviportal animals are large and broad. This configuration distributes the weight of the body over a large area. A larger object with more area can withstand a given force better than a smaller object with less area. It is the force per unit area that tests the strength of a structure.

Some mammals—such as kangaroos, many rodents, and some primates—specialized for bipedal jumping, and in these cases the distal segments of the hindlimb became lengthened. In jumping, the length of the hindlimb represents the distance through which the muscular force acts during jump takeoff. This distance is proportional to velocity of takeoff, which in turn determines height and distance of a jump. Long hindlimbs increase the takeoff velocity and the magnitude of the jump. In a few cases (galagos and tarsiers), the heel and ankle bones are also elongated, forming a fourth major limb segment to help propel the body forward rapidly. A few groups of mammals (gibbons and some New World monkeys) have limbs specialized for swinging arm over arm through the trees. This mode of locomotion, called brachiation, is correlated with long, hooklike hands and very long forelimbs for increased reach.

Other groups of mammals specialized their limbs for a high out-force. Members of these groups, including armadillos, anteaters, moles, and many rodents, are diggers. We can use simple lever mechanics to understand their common morphological features. Remember that the out-force is equal to (in-force × in-lever)/out-lever (chap. 7). Increasing the in-force and the in-lever and decreasing the out-lever are all ways to increase the forces of the hands and feet digging in the soil. The diggers are characterized by large muscle insertion scars on the limbs. These indicate an increase in the size of the muscles controlling limb movements relative to nonburrowers, or an increase in the in-force. Diggers also have relatively short limbs, or short out-levers. The olecranon process of most diggers is elongate. This process represents the in-lever length for the muscle controlling protraction of the lower arm. Lengthening this in-lever produces more out-force at the hand.

Mammals that specialized for aquatic life evolved long, streamlined bodies and shortened limbs. In intermediate aquatic specialists, such as seals and sea lions, the fore- and hindfeet were converted to paddles through the elongation of digits and the development of webbing. Extreme aquatic specialists such as whales and porpoises (cetaceans) and manatees and dugongs (sirenians) lost their hindlimbs, converted their forelimbs to flippers, and developed horizontal flukes on the tail fin that propels the animal through up-and-down power strokes. In contrast to fishes, aquatic amphibians and reptiles usually use lateral undulations of the body and tail fin for forward thrust. Also in contrast to amphibians and reptiles, no mammals have lost both fore- and hindlimbs to revert to a snakelike form.

Finally, one group of mammals, the bats (chiropterans), became specialized for active flight. The flight membrane of bats is composed of skin and supported by elongated forelimb bones, particularly the greatly elongated second to fourth digits. Unlike birds, bats did not develop an enlarged sternum for the attachment of flight muscles (and their flight muscles are not as large as those of birds), but they have some modifications of the shoulder joint for the downward flight propulsive stroke. The configuration of the flight surface is controlled by movements of the digits and by thin slips of muscles in the wing skin. Bats roost hanging upside down, and their hip and ankle joints are modified so extensively for hanging that quadrupedal locomotion is extremely clumsy, and bipedal locomotion is impossible.

The Evolutionary Radiations of the Mammals

When mammals began their explosive evolutionary radiations about 65–70 million years ago (December 25), the pattern of the radiations was strongly influenced by the positions of the continents. At that time, Eurasia, Greenland, and North America were connected across the North Atlantic, and they shared a common fauna. South America, Africa, and Australia were relatively isolated, and, for many millions of years, each major land mass had its own unique mammalian fauna.

In North America and Eurasia, a basal placental insectivore stock (order Proteutheria) radiated into a great diversity of small to large, archaic omnivores and herbivores, as well as archaic primates (order Primates, suborder Plesiadapiformes), archaic carnivores (order Creodonta), and the ancestors of modern carnivores (order Carnivora). In addition, there was a small diversity of opossum-like marsupials. Then, about 10 million years later, a second wave of evolutionary radiations in the northern continents produced the oldest members of the modern ungulate orders, the Artiodactyla (antelope, deer, camels, etc.) and Perissodactyla (horses, rhinos, tapirs, etc.), bats (order Chiroptera), modern Insectivora (shrews, moles, hedgehogs, etc.), pangolins (scaly anteaters, order Pholidota), and the first primates of modern aspect (suborder Prosimii). At about the same time in Asia, the first rodents (order Rodentia) and the first rabbits (order Lagomorpha) appeared. Archaic and modern types coexisted for 10–20 million years, and by the end of that time most of the marsupials and archaic placentals had become extinct. About 35 million years ago, there was a sharp climatic cooling episode and a wave of extinctions swept the northern continents. When it was over, many of the older families of the modern orders had become extinct, and a new wave of modern families of ungulates (particularly artiodactyls), rodents, and carnivores made their first appearance. The first higher primates (suborder Anthropoidea, the apes and monkeys) appeared in South America and Asia; apparently these animals migrated from northern continents.

163

In South America, a fossil record going back 70 million years shows an initial radiation that produced a great diversity of placental omnivores and herbivores, a fauna that coexisted with a modest diversity of marsupial insectivores, omnivores, and carnivores. Rodents and primates immigrated about 35 million years ago, and they diversified greatly but caused little apparent disruption among the native groups. Most of the indigenous South American mammal orders survived until about 2 million years ago, when a land bridge, the Isthmus of Panama, was established and linked South and North America. A great interchange of faunas took place across this bridge and in a relatively short period of time, which coincided with the beginning of the Pleistocene Ice Age, most of the native South American groups became extinct. Some of the native South American groups had been declining in diversity and number prior to the interchange, but the influx of new species from the north probably hastened their extinction. Modern survivors of this wave of extinctions include the marsupial opposum family and the placental order Edentata (sloths, armadillos, and anteaters). The long, 65-million-year isolation of the South American mammal fauna makes it a great natural experiment in convergent evolution. Carnivores and herbivores in South America independently evolved dental and locomotor specializations similar to those of their northern counterparts, even to the extremes represented by sabertoothed marsupial carnivores and one-toed horselike ungulates!

Although the Cenozoic fossil record of African mammals extends back 55 million years, that record is very sparse for the first half of the Age of Mammals. Our first good fossil record already includes emigrants from Eurasia, such as primates, rodents, archaic carnivores, and artiodactyls. However, it also includes some indigenous African forms—ancestral elephants (order Proboscidea), hyraxes (order Hyracoidea), and elephant shrews (order Macroscelidea). Elephants later made their way onto nearly every continent, but Macroscelidea have been confined to Africa and the hyraxes to Africa and adjacent parts of Asia throughout their histories. An unusual order of mammals that is now indigenous to Africa but that may have originated in Europe is the aardvarks (order Tubulidentata), which are peg-toothed, powerfully built termite- and anteaters.

Australia had a very different evolutionary history, because for most of the Age of Mammals its mammalian fauna consisted only of monotremes, marsupials, and their fossil allies. Marsupials filled virtually all the ecological niches occupied by placentals on other continents and included animals that were insectivores, carnivores, omnivores, and herbivores. Marsupials never evolved swift runners, like the placental ungulates, but the large leaping kangaroos are locomotor analogues. There even were marsupial anteaters and marsupial moles. Many of the large marsupials have become extinct within the last million years, at a time when large mammals became extinct on other continents as well, and some of the most unusual of the marsupial carnivores and ungulate equivalents, such as the jaguar-sized *Thylacoleo* or the rhino-sized *Diprotodon,* no longer exist. Australia was a refuge not only for marsupials but also for the duck-billed platypus and echidna, the only surviving egg-laying mammals. As mentioned in chapter 16, these unusual animals represent a group that diverged from the common ancestor of marsupials and placentals about 180 million years ago, shortly after the origin of mammals, and they are placed in an order of their own, the Monotremata.

Our very brief survey of mammal radiations would not be complete without consideration of the marine mammals. Seals and sea lions evolved about 25 million years ago (December 30) from two different groups of land carnivores, the mustelids (otter family) and ursids (bears), respectively. Far more ancient are the orders Cetacea (whales and porpoises) and Sirenia (manatees and dugongs), which first appeared in the oceans about 50 million years ago.

The approximately 4,000 species of living mammals are classified into twenty-one orders. The three largest orders are rodents (about 40% of the total number of living mammalian species), bats (about 20%), and insectivorans (about 10%).

The modern mammal that is perhaps most interesting to us is *Homo sapiens,* and therefore we will spend some time on our own evolutionary history. We are members of the order Primates, and primate evolutionary history can be viewed in terms of three major evolutionary radiations (fig. 18.1). The earliest primates, suborder Plesiadapiformes, appeared about 65 million years ago (December 25), at the beginning of the Age of Mammals. They had a pair of enlarged, superficially rodent-like incisor teeth, large olfactory bulbs, small eyes, and large vibrissae whose presence we infer from fossil records of the canal for their nerves and blood vessels. Most were probably nocturnal and insectivorous, like modern shrews and their relatives. They underwent a modest evolutionary radiation of mostly small (mouse- to squirrel-sized) animals, which looked very different from modern primates.

Most plesidapiforms became extinct about 55 million years ago,

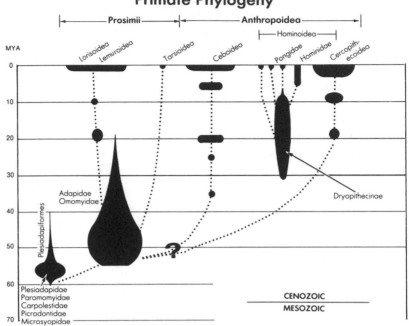

Fig. 18.1   Primate phylogeny and geological ranges of primate groups.

when the second primate evolutionary radiation began. This second radiation produced the oldest members of the suborder Prosimii, which includes the modern lemurs, lorises, galagos, and tarsiers. The first prosimians looked a lot like modern ones, except that they had brains about half the size (relative to body size) of those we see in modern prosimians (fig. 18.2). They ranged from mouse sized to cat sized, and some had enlarged, pointed incisor teeth, while others had small spatulate ones. Fossil brain casts indicate that the visual and possibly the auditory areas of the brain were enlarged and the olfactory bulbs were reduced compared to the primitive condition, changes suggesting that vision and possibly hearing took on increased importance while olfaction became less significant.

Rare skeletal remains indicate that some of the early prosimians had locomotor specializations for leaping—elongate hindlimbs and elongated heel and ankle bones—like the modern galagos and tarsiers, while others were arboreal quadrupeds. For 20 million years, early prosimians were common elements in mammal faunas in North America

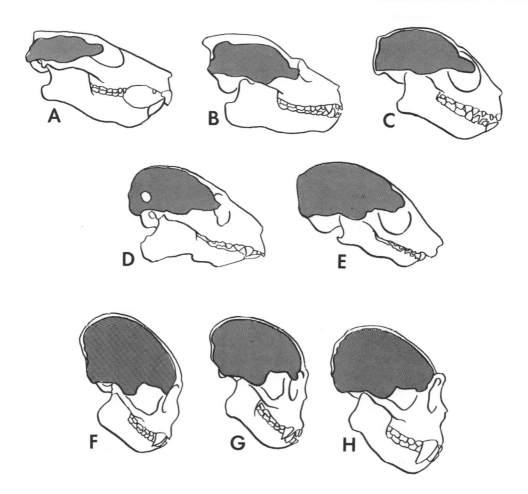

Fig. 18.2   Diversity of skull shape and brain size in Primates. *A, Plesiadapis,* a plesiadapid (after Szalay and Delson, 1979). *B, Adapis,* an adapid (after Szalay and Delson, 1979). *C, Tetonius,* an omomyiid (after Szalay and Delson, 1979). *D, Indri,* a prosimian. *E, Microcebus,* a prosimian. *F, Cebus,* a ceboid. *G, Cercopithecus,* a ceropithecoid. *H, Hylobates,* a hominoid.

and Europe. The smaller ones were probably insectivorous and the larger ones herbivorous, much like modern prosimians, and, judging from orbit sizes, some were nocturnal and others diurnal.

Then, about 35 million years ago, the last of the plesiadapiforms and most of the prosimians became extinct in the great wave of extinctions that wiped out many other groups of mammals on northern continents. Some prosimians survived to give rise to the lemurs, lorises, and tarsiers found today in Africa and Asia and to the monkeys, apes, and humans (suborder Anthropoidea), which constitute the third evolutionary radiation of the primates.

The oldest known anthropoids are about 30 million years old (December 29), and include the first South American monkeys and the oldest apes, the latter known from Africa (specifically, Egypt). Anthropoids have more powerfully built jaws and relatively larger brains than prosimians and also more expanded visual areas of the brain and more reduced olfactory bulbs (fig. 18.2). South American monkeys (superfamily Ceboidea) had a modest radiation, resulting in the approximately seventy-five living species, ranging from tiny insectivorous pygmy marmosets to the medium-sized fruit- and leaf-eating wooly, spider, and howler monkeys. (The latter three are the only monkeys with prehensile, i.e., gripping tails.) Old World apes (superfamily Hominoidea) underwent a small evolutionary radiation between about 25 and 10 million years ago. They were gibbon- to chimp-sized, arboreal, and primarily fruit eaters, but without the locomotor specializations of modern apes (brachiating and knuckle walking), and they spread frm Africa to Europe and Asia. About 20 million years ago, Old World monkeys (Cercopithecoidea in fig. 18.1) evolved from these early apes, and about 10 million years ago they underwent an evolutionary radiation that produced the sixty species of modern African and Asian monkeys, ranging from the small talapoins through medium-sized macaques and langurs, to the large, ground-living baboons and their relatives. The radiation of cercopithecoid monkeys coincided with the demise of most of the apes, and, although we are accustomed to thinking of apes as more "advanced" than monkeys, it is likely that the latter largely outcompeted the apes as arboreal herbivores.

Some of the primitive apes survived to give rise to the modern Asiatic gibbons and orangutans, the modern African chimps and gorillas, and an animal that first appeared about 4 million years ago, the terrestrial bipedal form called *Australopithecus,* the oldest member of our own family, the Hominidae. The oldest species of *Australopithecus* stood about 4 feet tall. It had a brain slightly larger than that of modern monkeys and apes and relatively small canine and incisor teeth, but otherwise its cranial features were generally similar to those of its ape ancestors (fig. 18.3). Where it differed dramatically from any ape, or any other primate for that matter, was in its locomotor system, which, as far back as 4 million years ago, was modified for erect bipedal loco-

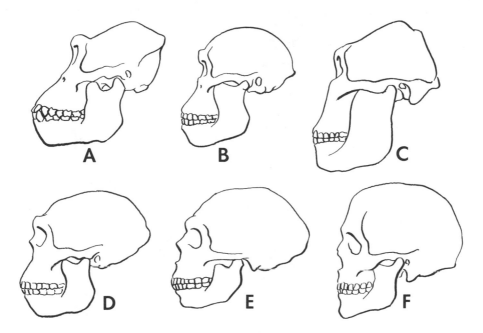

Fig. 18.3   Hominoid skull diversity. *A*, female gorilla (after Campbell, 1979). *B, Australopithecus africanus* (after Campbell, 1979). *C, Australopithecus robustus. D, Homo erectus* (after Howells, 1979). *E, Homo sapiens* (neanderthal) (after Campbell, 1979). *F, Homo sapiens,* modern man (after Campbell, 1979).

motion. This modification involved not only changes in the hindlimbs, including drastic remodeling of the pelvis, knees, and feet, but also remodeling of the trunk to facilitate habitual balancing in an erect posture and changes in the articulation of the head upon an erect vertebral column (fig. 18.4). *Australopithecus* evolved into two lineages. One, *A. robustus,* grew to a larger size and developed a massive jaw apparatus with small anterior teeth and large grinding molars. The other, *A. africanus,* remained relatively small and slender, with a less specialized jaw apparatus. The large, robust *Australopithecus* survived until about 1 million years ago. The smaller *A. africanus* gave rise, by 1.8 million years ago, to the oldest members of our own genus, *Homo habilis* and *Homo erectus.*

The main changes involved in evolution from *Australopithecus* to *Homo* were a small increase in body size and a significant increase in relative brain size. *Homo erectus* persisted for over a million years, made crude stone tools (a significant cultural advance over *Aus-*

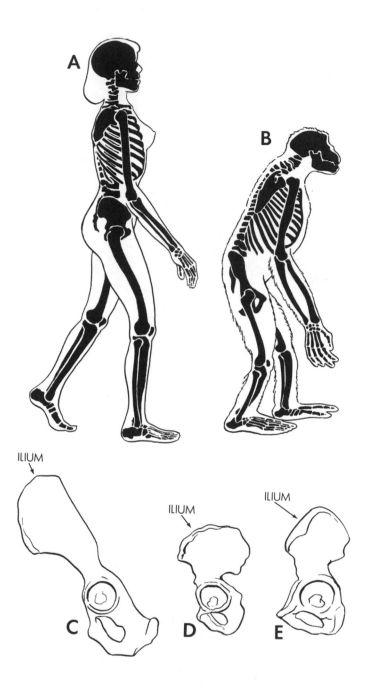

Fig. 18.4 *A*, female human (after Napier, 1979). *B*, male chimpanzee (after Clark, 1971). *C*, the pelvis of a gorilla (after Napier, 1979). *D*, the pelvis of *Australopithecus* (after Napier, 1979). *E*, the pelvis of *Homo* (after Napier, 1979).

tralopithecus), hunted a variety of small and large game, and, at least during the latter part of its time span, had the use of fire. Somewhere between 125,000 and 250,000 years ago, *Homo erectus* evolved into modern *Homo sapiens*. The main change was another significant increase in relative brain size that brought us to the modern human range, in which the brain is about 3.5 times as large as the brains of monkeys or apes. This increase in brain size brought about dramatic changes in the shape of our skulls. Although it is commonly believed that bigger brains mean greater intelligence, there is little scientific evidence for this belief. At the present time we have not solved the enigma of why our uniquely enlarged brain evolved.

Between about 75,000 and 35,000 years ago, we have a substantial record of a race of *Homo sapiens*, known as Neanderthals, that lived mainly in Europe. They stood about 5.5 feet tall and were more powerfully built than modern humans. Neanderthals hunted a great variety of game animals with more sophisticated stone and bone tools that *H. erectus* had. About 30,000 to 35,000 years ago, Neanderthals were succeeded by people whose skeletons are indistinguishable from those of modern humans and who are called Cro-Magnons. Cro-Magnons had more sophisticated stone and bone tools and produced the first known works of art. Tools similar to Cro-Magnon's appear in other parts of the world at about the same time, and this coincidence suggests some sort of breakthrough in hunting technology. A major technological/cultural revolution, the Neolithic Revolution, began about 10,000 years ago, with the first domestication of plants and animals, and the increase in the size of the human population, which is now reaching crisis proportions, began then. Cultural evolution had become significant by at least the time of *Homo erectus,* as indicated by the stone tool technology, but, judging from the sterotypical nature of the stone tools, the manufacturing technique appears to have remained relatively static for the million years of *Homo erectus*'s history. On the other hand, in the 10,000 years since the Neolithic Revolution, cultural and technological changes have been proceeding at an exponential rate, with a truly incredible explosion of technological advances just in the past 100 years. We now have the technological ability to do what no other animal has been able to accomplish. We can alter our environments and thus alter the conditions and forces that, until this point in the history of life, have driven natural selection and the evolutionary process. Among our abilities now is the potential to render ourselves and most other species on Earth extinct. The next few decades should determine the outcome.

Glossary

ABDUCT.   To move a part away from the body in the sagittal plane or to move two parts away from each other.

ADDUCT.   To move a part toward the sagittal plane of the body or move two parts together.

ALLOMETRY.   The study of changes of shape with size.

AMPLITUDE.   The range of movement or displacement.

ANTAGONIST.   A muscle that opposes the action of another.

ARCHAIC.   Oldest members of a lineage.

ARTIODACTYL.   An order of even-toed, hooved mammals including deer, antelope, pigs, and goats.

BALLAST.   A structure that gives stability.

BARBULE.   Small projections that interlock on a feather.

BIOMECHANICAL COUPLE.   Two equal, parallel forces acting on an object in opposite directions, causing rotation.

CARNIVORE.   Meat-eating.

CENTRUM.   The body of a vertebra.

COUNTERWEIGHT.   An equivalent weight.

CRETACEOUS.   A geological time period from 135 to 60 million years ago.

DENTICLE.   Small, toothlike structure.

DENTINE.   A tissue of teeth and scales, softer than enamel but harder than bone.

DISTAL.   Farther from the center of the body.

DORSAL.   On the upper, back, or vertebral side of the body.

DRAG.   The resistance of a body to movement through air or water.

ECTODERM.   The outermost of the three embryonic tissues.

ECTOTHERM.   Using a source of heat outside the body.

ENDOTHERM.   Using a source of heat inside the body.

EXTENSOR.   A muscle that, when contracting, stretches the limb.

FLEXOR.   A muscle that, when contracting, bends the limb.

FORCE.   A push or pull that causes motion, mass × acceleration.

GENOTYPE.   Genes of an individual.

GRAVIPORTAL.   Heavy bodied animals.

HERBIVORE.   Plant-eating.

HETEROCERCAL TAIL.   An asymmetrical tail with the notochord extending into the dorsal lobe.

173

HOMOCERCAL TAIL.   A symmetrical tail.

HYOMANDIBULA.   The major upper segment of the hyoid gill arch.

IMPEDANCE.   The ratio of the pressure displacement to the volume displacement at a given surface.

INSECTIVORE.   Insect-eating.

ISCHIUM.   A bone of the pelvic girdle.

KERATIN.   A hard, protein material.

KINETIC ENERGY.   The energy of motion.

LIFT.   The force that is at right angles to the oncoming air or water.

LIGAMENT.   Cords or sheets of connective tissue joining bones.

LOAD.   A force applied to a solid object.

MESODERM.   The middle embryonic tissue.

METAPODIAL.   Collective term for the metacarpals and metatarsals.

MOMENT OF INERTIA.   The tendency of an object to resist movement.

MYOMERE.   Block of muscle of a body segment.

MYOTOME.   Muscle blocks.

NEURAL TUBE.   The embryonic central nervous system.

NOTOCHORD.   Internal support rod for the body, present at some stage in the life of a vertebrate.

ONTOGENY.   The embryonic development of an individual.

OPERCULUM.   The flap covering the gills of some fishes.

PERILYMPHATIC FLUID.   The clear fluid of the inner ear.

PHALANGES.   Finger and toe bones.

PHENOTYPE.   Visible characters of an individual.

PHYLOGENY.   Evolutionary history of an organism.

PLACENTA.   An organ of embryonic and maternal tissues that functions in the physiological exchange of food, gas, and wastes.

POLARITY.   Direction of modification of a character.

POTENTIAL ENERGY.   The energy of position.

POWER.   The rate of doing work, force $\times$ distance $\times$ velocity.

PROTRACT.   Reach out from the body in the sagittal plane.

PROXIMAL.   Closer to the center of the body.

RETRACT.   Pull back toward the body in the sagittal plane.

SACRAL.   The area of the vertebral column where the pelvic girdle attaches.

SAGITTAL.   A plane that divides the body into right and left sides.

STERNUM.   Breast bone.

TETRAPOD.   Four feet, usually referring to land vertebrates.

THRUST.   A force.

TORQUE.   A turning force.

TRANSVERSE.   A plane that divides the body into anterior and posterior parts.

UNGULATE.   Animals with only hooves in contact with the ground when walking, includes horses, deer, antelopes.

VENTRAL.   On the lower, belly, or underside of the body.

ZYGAPOPHYSES.   Interlocking processes on the vertebrae of land tetrapods.

Additional Readings

Alexander, R. 1981. *The chordates*. 2d ed. London: Cambridge University Press.

Colbert, E. 1980. *Evolution of the vertebrates*. 3d ed. New York: John Wiley & Sons.

Hildebrand, M. 1982. *Analysis of vertebrate structure*. 2d ed. New York: John Wiley & Sons.

Hildebrand, M., D. Bramble, K. Liem, and D. Wake, 1985. *Functional vertebrate morphology*. Cambridge, Mass.: Harvard University Press.

Piveteau, J., ed. 1964. *Traite de paleontologie*. Paris: Masson.

Romer, A. S. 1966. *Vertebrate paleontology*. Chicago: University of Chicago Press.

Chapter 2

Patterson, C. 1978. *Evolution*. Ithaca, N.Y.: Cornell University Press.

Smith, J. 1982. *Evolution now*. San Francisco: W. H. Freeman & Co.

Chapter 4

Halstead, L. B. 1973. The heterostracan fishes. *Biol. Rev.* 48:279–332.

Janvier, P. 1984. The relationships of the osteostraci and galeaspids. *J. Vert. Paleo.* 4:344–58.

Moy-Thomas, J., and R. Miles. 1971. *Palaeozoic fishes*. 2d ed. Philadelphia: W. B. Saunders Co.

Chapter 5

Forey, P. 1984. Yet more reflections on agnathan-gnathostome relationships. *J. Vert. Paleo.* 4:330–43.

Mallatt, J. 1984. Early vertebrate evolution: Pharyngeal structures and the origin of gnathostomes. *J. Zool. Lond.* 204:169–81.

Chapter 6

Pearson, D. M. 1982. Primitive bony fishes, with special reference to *Cheirolepis* and palaeonisaform actinoptergians. *Zool. J. Linn. Soc.* 74:35–67.

Chapter 7

Thomson, K., and D. Simanek. 1977. Body form and locomotion in sharks. *Amer. Zool.* 17:343–54.

Chapter 8

Lauder, G., and K. Liem. 1983. The evolution and interrelationships of the actinopterygian fishes. *Bull. Mus. Comp. Zool.* 150:95–197.

Webb, P. 1982. Locomotor patterns in the evolution of actinopterygian fishes. *Amer. Zool.* 22:329–42.

Chapters 9 and 10

Edwards, J. 1977. The evolution of terrestrial locomotion. In M. Hecht, et al., eds. *Major patterns in vertebrate evolution.* New York: Plenum Press.

Lombard, R., and J. Bolt. 1979. Evolution of the tetrapod ear: An analysis and reinterpretation. *Biol. J. Linn. Soc.* 11:19–76.

Panchen, A., ed. 1980. *The terrestrial environment and the origin of land vertebrates.* New York: Academic Press.

Chapter 11

Holmes, R. 1977. The osteology and musculature of the pectoral girdle of small captorhinids. *J. Morphol.* 152:101–40.

Reisz, R. 1981. A diapsid reptile from the pennsylvanian of Kansas. Special Publication. *Mus. Nat. Hist. Univ. Kansas,* no. 7, 1–74.

Chapter 12

Charig, A. 1979. *A new look at the dinosaurs.* London: Heinemann.

Norman, D. 1985. *The illustrated encyclopedia of dinosaurs.* New Haven, Conn.: Crescent Books.

Thomas, R., and E. Olsen, eds. *A cold look at the warm blooded dinosaurs.* AAAS Selected Symposium no. 28. Boulder, Colo.: Westview Press.

Chapter 14

Carroll, R. 1985. Evolutionary constraints in aquatic diapsid reptiles. *Special papers in paleontology,* no. 33, 145–55.

Russell, D. 1967. Systematics and morphology of American Mosasaurs. *Peabody Mus. Nat. Hist. Yale Univ. Bull.* 23:1–241.

Chapter 15

Hecht, M., J. Ostrom, G. Viohl, and P. Wellnhofer, eds. 1985. *The beginnings of birds.* Eichstatt: Bronner & Daentler.

Padian, K. 1983. A functional analysis of flying and walking in pterosaurs. *Paleobiology.* 9:218–39.

Chapter 16

Crompton, A., and P. Parker. 1978. Evolution of the mammalian masticatory apparatus. *Amer. Sci.* 62:192–201.

Kemp, T. 1982. *Mammal-like reptiles and the origin of mammals.* London: Academic Press.

Chapter 17

Radinsky, L. 1985. Patterns in the evolution of ungulate jaw shape. *Amer. Zool.* 25:303–14.

Smith, J., and R. Savage. 1956. Some locomotory adaptations in mammals. *Zool. J. Linn. Soc.* 42:603–22.

Chapter 18

Pfeiffer, J. 1985. *The emergence of man.* New York: Harper & Row.

Radinsky, L. 1975. Primate brain evolution. *Amer. Sci.* 63:656–63.

Figure Credits

All of the illustrations were prepared by Dennis Green. Some of these were adapted from illustrations prepared by others, as indicated below:

Fig. 1.1. Jacques Gauthier, unpublished.

Fig. 4.1. *A:* Stensio, E. 1939. A new anaspid from the upper Devonian of Scaumenac Bay in Canada, with remarks on the other anaspids. *Stockholm K. Vet. Akad. Handl.* 18:1–25. Piveteau, J., ed. 1964. *Traite de paleontologie* IV Vol 1 Agnathes. Paris: Masson et Cⁱᵉ.

C: Janvier, P. 1977. Contribution a la connaissance de la systematique et du l'anatomie du qenre *Boreaspis* Stensio (Agnatha Cephalaspidomorphi, Osteostraci) du Devonien inferieur du Spitsberg. *Ann. Paleont.* 63:1–32.

D and *E:* Stensio, E. 1932. *The Cephalaspids of Great Britain.* London: British Museum (Natural History).

Fig. 4.2. *A:* Piveteau, J., ed. 1964. *Traite de paleontologie IV,* vol. 1. Agnathes. Paris: Masson et Cⁱᵉ.

B: Heintz, A. 1958. The head of the anaspid *Birkenia elegans* Traquair. In *Studies on fossil vertebrates presented to D. M. S. Watson,* edited by T. S. Westoll, 71–85. London: Athcone Press.

Fig. 4.3. *A:* White, E. 1935. The ostracoderm *Pteraspis kner* and the relationships of the agnathous vertebrates. *Phil. Trans. R. Soc.* ser. B, 225:381–457.

C: Heintz, N. 1968. The pteraspid *Lyktaspis* n.g. from the Devonian at Vestspitsbergen. *Nobel Symposium* 4:73–80.

D: Obruchev, D. 1967. *Fundamentals of paleontology.* vol. 2, Agnatha, Pisces. Jerusalem: Israel Program for Scientific Translations.

E: Obruchev, D. 1943. A new restoration of *Drepanaspis*. U.S.S.R. *Akad Sci C. R.* (Doklady) 41:268–71.

Fig. 5.2. *A* and *B:* Waston, D. 1937. The acanthodian fishes. *Phil. Trans. R. Soc.,* Ser. B, 228:49–146.

C: Miles, R. 1966. The acanthodian fishes of the Devonian Plattenkalk of the Puffrath Trough in the Rhineland, with an appendix containing a classification of the Acanthodii and a revision of the genus *Homalacanthus*. *Ark. Zool.* 18:147–94.

Fig. 5.3. *A* and *B:* Miles, R., and T. Westoll. 1968. The placoderm fish *Coccosteus cuspidatus* Miller ex Agassiz from the Middle Old Red Sand-

179

stone of Scotland. Pt. 1, Descriptive morphology. *Trans. R. Soc. Edinb.* 67:373–476.

C: Stensio, E. 1963. Anatomical studies on the arthrodiran head. Pt. 1, Preface, geological and geographical distribution, the organization of the arthrodires, the anatomy of the head in the Dolichothoraci, Coccosteomorphi, and Pachyosteomorphi. *K. Svenska Vetensk Akad. Handl.* 9:1–419.

D and *E:* Piveteau, J., ed. 1969. *Traite de Paleontologie IV,* vol. 11. Gnathostomes, Acanthodiens, Placodermes, Elasmobranches. Paris: Masson et Cie.

Fig. 6.1.   *A:* Moy-Thomas, J., and R. Miles. 1971. *Palaeozoic fishes.* 2d ed. Philadelphia: W. B. Saunders Co.

B and C: Colbert, E. 1980. *Evolution of the vertebrates.* 3d ed. New York: John Wiley & Sons.

D–F: Pearson, D. 1982. Primitive bony fishes, with special reference to *Cheirolepis* and palaeonisciform actinopterygians. *Zool. J. Linn. Soc.* 74:35–67.

Fig. 7.1.   *A:* Moy-Thomas, J., and R. Miles. 1971. *Palaeozoic fishes.* 2d ed. Philadelphia: W. B. Saunders Co.

Fig. 7.2.   *A* and *B:* Thomson, K., and D. Simanek. 1977. Body form and locomotion in sharks. *Amer. Zool.* 17:343–54.

C–G: Gregory, W. K. 1951. *Evolution emerging,* vol 2. New York: Macmillan.

Fig. 8.3.   Liem, K. 1978. Modulatory multiplicity in the functional repertoire of the feeding mechanism of cichlid fishes. I. Piscivores. *J. Morphol.* 158:323–60.

Fig. 8.4.   Gregory, W. K. 1959. *Fish skulls.* New York: Noble Offset Printers.

Fig. 8.2.   Lindsey, C. 1978. Form, function, and locomotory habits. In *Fish physiology,* vol. 7, *Locomotion,* edited by W. Hoar and D. Randall, 1–100. New York: Academic Press. Webb, P., and R. Blake. 1985. Swimming. In *Functional vertebrate morphology,* edited by M. Hildebrand, D. Bramble, K. Liem, and D. Wake, 110–128. Cambridge, Mass.: Harvard University Press. Young, J. Z. 1962. *The life of vertebrates.* 2d ed. London: Oxford University Press.

Fig. 9.3.   Gregory, W. 1951. *Evolution emerging.* vol. 2. New York: Macmillan. Andrews, S., and T. Westoll 1970. The postcranial skeleton of *Eusthenopteron foordi* Whiteaves. *Trans. Roy. Soc. Edinb.* 68:207–329.

Fig. 10.2.   *A, C–F:* Romer, A. S. 1966. *Vertebrate paleontology.* 3d ed. Chicago: University of Chicago Press.

B: Gregory, W. 1951. *Evolution emerging,* vol. 2. New York: Macmillan.

Fig. 10.3.   *A:* Carroll, R. 1968. The postcranial skeleton of the Permian microsaur *Pantylus. Can. J. Zool.* 46:1175–92.

B: Milner, A. 1980. A review of the Nectridea (Amphibia). In *The terrestrial environment and the origin of land vertebrates,* edited by A. Panchen, 377–405. London: Academic Press.

C and *D:* Romer, A. S. 1966. *Vertebrate paleontology.* 3d ed. Chicago: University of Chicago Press.

D: Beerbower, J. 1963. Morphology, paleoecology and phylogeny of the Permo Pennsylvanian amphibian *Diploceraspis. Bull. Mus. Comp. Zool.* 130:31–108.

Fig. 10.4.   *A:* Schaeffer, B. 1941. The morphological and functional evolution of the tarsus in amphibians and reptiles. *Bull. Amer. Mus. Nat. Hist.* 78:395–472.

B: Gans, C. 1974. *Biomechanics.* Philadelphia: J. Lippincott Co.

Fig. 11.2.   *A:* Carroll, R. 1969. Problems of the origin of reptiles. *Biol. Rev.* 44:151–70.

B: Gregory, W. 1951. *Evolution emerging,* vol. 2. New York: Macmillan.

C: Romer, A., and L. Price. 1940. Review of the Pelycosauria. *Geol. Soc. Am. Spec. Pap.* 28:1–538.

D: Romer, A. S. 1966. *Vertebrate paleontology.* 3d ed. Chicago: University of Chicago Press.

Fig. 11.3.   *A:* Heaton, M., and R. Reisz. 1980. A skeletal reconstruction of the early Permian captorninid reptile *Eocaptorhinus laticeps* (Wiliston). *J. Paleo.* 54:136–43.

B and *C:* Heaton, M. 1980. The Cotylosauria: A reconsideration of a group of archaic tetrapods. In *The terrestrial environment and the origin of land vertebrates,* edited by A. Panchen, 497–551. London: Academic Press.

D: Romer, A. S. 1966. *Vertebrate paleontology.* 3d ed. Chicago: University of Chicago Press.

Fig. 11.4.   *A:* Barghusen, H. 1975. A review of fighting adaptations in dinocephalians (Reptilia, Therapsida). *Paleobiology* 1:295–311. Gregory, W. 1951. *Evolution emerging,* vol. 2. New York: Macmillan.

B: Crompton, A. W. 1972. The evolution of the jaw articulation of cynodonts. In *Studies in vertebrate evolution,* edited by K. Joysey and T. Kemp, 231–53. Edinburgh: Oliver & Boyd.

C: Romer, A. S. 1966. *Vertebrate paleontology.* 3d ed. Chicago: University of Chicago Press.

D: Kemp, T. 1982. *Mammal-like reptiles and the origin of mammals.* London: Academic Press.

Fig. 12.2.   *A:* Camp, C. 1930. A study of phytosaurs. *Mem. Univ. California* 10:1–161.

B: Colbert, E. 1980. *Evolution of the vertebrates.* 3d ed. New York: John Wiley & Sons.

Fig. 12.4.   *A:* Ewer, R. 1965. The anatomy of the thecodont reptile *Euparkeria. Phil. Trans. Roy. Soc.* London, Ser. B, 248:379–435.

B and *F:* Romer, A. S. 1966. *Vertebrate paleontology.* 3d ed. Chicago: University of Chicago Press.

C and *E:* Gregory, W. 1951. *Evolution emerging,* vol. 2. New York: Macmillan.

D: Ostram, J. 1961. Cranial morphology of the hadrosaurian dinosaurs of North America. *Bull. Amer. Mus. Nat. Hist.* 122:33–186.

Fig. 12.5.   *A:* Colbert, E. 1980. *Evolution of the vertebrates.* 3d ed. New York: John Wiley & Sons. Charig, A. 1979. *A new look at the dinosaurs.* London: Heinemann.

B: Gregory, W. 1951. *Evolution emerging,* vol. 2. New York: Macmillan.

C: Gilmore, C. 1936. Osteology of *Apatosaurus* with special reference to specimens in the Carnegie Museum. *Mem. Carnegie Mus.* 11:175–300.

D: Romer, A. S. 1966. *Vertebrate paleontology.* 3d ed. Chicago: University of Chicago Press.

E: Gilmore, C. 1925. A nearly complete articulated skeleton of *Camarasaurus,* a saurischian dinosaur from the Dinosaur National Monument, Utah. *Mem. Carnegie Mus.* 10:347–84.

F: Young, J. 1962. *Life of vertebrates.* 2d ed. London: Oxford University Press. Bakker, R. 1980. Dinosaur heresy-dinosaur renaissance: Why we need endothermic archosaurs for a comprehensive theory of bioenergetic evolution. In *A cold look at the warm-blooded dinosaurs,* edited by M. D. Thomas and E. Olson, 351–505. Boulder, Colo.: Westview Press.

Fig. 13.1.   Romer, A. S. 1956. *Osteology of the reptiles.* Chicago: University of Chicago Press. Gregory, W. 1951. *Evolution emerging,* vol. 2. New York: Macmillan. Huene, F. 1956. *Palaontologie und phylogenie der niederen tetrapoden.* Jena: Fischer.

Fig. 13.2.   Romer, A. S. 1956. *Osteology of the reptiles.* Chicago: University of Chicago Press. Colbert, E. 1980. *Evolution of the vertebrates.* 3d ed. New York: John Wiley & Sons. Smith, M. 1943. *The fauna of British India: Reptilia and Amphibia,ˆ* vol. 3, *Serpentes.* London: Taylor, and Francis. Frazetta, T. 1962. A functional consideration of cranial kinesis in lizards. *J. Morphology* 111:287–320.

Fig. 14.1.   Romer, A. S. 1966. *Vertebrate paleontology.* 3d ed. Chicago: University of Chicago Press. Gregory, W. 1951. *Evolution emerging,* vol. 2. New York: Macmillan. Russell, D. 1967. Systematics and morphology of American mosasaurs. *Peabody Mus. Nat. Hist. Yale Univ. Bull.* 23:1–241.

Fig. 15.1.   Padian, K. 1983. A functional analysis of flying and walking in pterosaurs. *Paleobiology* 9:218–39. Eaton, G. 1910. Osteology of *Pteranodon. Mem. Connecticut Acad. Sci.* 2:1–38.

Fig. 15.3.   Hildebrand, M. 1982. *Analysis of vertebrate structure.* 2d ed. New York: John Wiley & Sons.

Fig. 16.2.   Crompton, A., and P. Parker. 1978. Evolution of the mammalian masticatory apparatus. *Amer. Sci.* 66:192–201. J. Hopson, unpublished.

Fig. 16.3.   Barghusen, H., and J. Hopson. 1979. The endoskeleton: The comparative anatomy of the skull and visceral skeleton. In *Hyman's comparative vertebrate anatomy,* edited by Marvalee Wake. 3d ed. Chicago: University of Chicago Press. J. Hopson, unpublished.

Fig. 17.4.   Gregory, W. 1951. *Evolution emerging,* vol. 2. New York: Macmillan. Young, J. 1962. *The life of vertebrates.* 2d ed. London: Oxford University Press.

Fig. 18.2.   *A–C:* Szalay, F., and E. Delson. 1979. *Evolutionary history of the primates.* New York: Academic Press.

Fig. 18.3.   Howells, W. 1979. *Homo erectus.* In *Human ancestors,* with introduction by G. Issac and R. Leaky. San Francisco: W. H. Freeman. Campbell, B., ed. 1979. *Humankind emerging.* Boston: Little, Brown.
Fig. 18.4.   Clark, W. Le Gros. 1971. *The antecedents of man.* 3d ed. Chicago:
Quadrangle Books. Napier, J. 1979. The antiquity of human walking. In
Human ancestors, with introduction by G. Issac and R. Leaky. San Francisco: W. H. Freeman.

Index